Propagators for many-particle systems

Propagators for many-particle systems

An elementary treatment

ROBERT MILLS
The Ohio State University

GORDON AND BREACH SCIENCE PUBLISHERS
New York London Paris

Library of Congress catalog card number: 69-16985

Editorial office for the United Kingdom:

Gordon and Breach Science Publishers Ltd.
12 Bloomsbury Way
London W. C. 1

Editorial office for France:

Gordon and Breach
7-9 rue Emile Dubois
Paris 14e

Distributed in Canada by:

The Ryerson Press
299 Queen Street West
Toronto 2B, Ontario

To Lee

Preface

IT IS the purpose of this book to provide a fairly straightforward and unsophisticated introduction to the field-theory approach in the many-body problem, suitable for nontheoretical students and others who wish to familiarize themselves with the terminology and basic ideas in this area. In order to give greater emphasis to the more fundamental aspects of the subject, a number of important topics, well-treated elsewhere, are not included here. There is no discussion, for example, of the quasiparticle approximations, the self-consistent ladder approximation, the random phase approximation, or the theory of superconductivity, to name a few. It is hoped that the present treatment will provide a suitable background for further study in these areas.

The development centers on the single-particle propagator, or Green's function, which describes the behavior of the system when one particle is added to or removed from an equilibrium state, and which can be used to determine the macroscopic properties of the system. This propagator takes a variety of forms, the general properties and relationships of which are discussed in detail. The perturbation expansion procedure and Feynman graph prescription are worked out, both for the ground state and for the case of finite temperatures. A number of the properties of these propagators are summarized for convenience in an appendix, together with a concise statement of the various Feynman graph prescriptions developed in the text.

Some of the results used in this book came out of research supported in part by the National Science Foundation and in part by the Department of Scientific and Industrial Research of the Government of Great Britain, whose assistance is most gratefully acknowledged.

Robert Mills

Contents

1

Introduction

IT SHOULD be stressed at the very beginning that the use of field theory in dealing with many-particle systems is almost exclusively a matter of mathematical convenience, being, except in rare instances, mathematically equivalent to the use of the many-particle Schrödinger equation. The form of the equations is very different in the field theory approach, however, and in many respects much simpler, so that methods of expansion and approximation suggest themselves which would be very difficult even to formulate using the Schrödinger equation. Also, some of the basic properties of the system, such as the behavior of the elementary excitations, can be expressed and analyzed readily in terms of the functions used in the field-theory approach, while they would be much harder to pick out from the properties of the wave functions in the Schrödinger picture.

The basic idea is simply this: when one applies the rules of quantization to a field, (consider the electromagnetic field as an example), then one finds that the excited states of the system behave in every way as if particles were present, (photons, in the electromagnetic case), whose properties are determined by the properties of the original field. This happens in the following way: for the noninteracting field, each normal mode of the field behaves like a linear harmonic oscillator, and is quantized in the same way, giving rise to equally spaced energy levels with spacing $h\nu_c$ where ν_c is the classical frequency of that normal mode. If this normal mode can be taken as a running wave (using periodic boundary conditions in a rectangular box, say), then the excited states of the normal mode, regarded as an oscillator, will also be eigenstates of the total momentum of the field with equal spacings of magnitude h/λ, where λ is the wavelength of the normal mode. Thus the energy and momentum of the system, if this normal mode is in its nth excited state, are the same as if n particles were present, each of energy $h\nu_c$ and momentum

h/λ. This equivalence is complete, and permits us, if we wish, to regard all particles as the quanta of appropriate fields; the properties of the fields are chosen to give the quanta the proper properties (mass, spin, etc.), since except for the electromagnetic and gravitational cases the field itself is not observed as such.

It seems to be necessary to make this identification when the particles can be produced or absorbed in physical processes, since there the wave function description breaks down, or is at best very cumbersome indeed. At relativistic energies this is true of all particles, so that it is widely felt that, within the framework of quantum mechanics, field theory, or some improvement on field theory, provides the only possible basic description of nature. For nonrelativistic problems in which no production or absorption processes take place, the field theory description becomes equivalent to a wave function approach, as we have mentioned, and its use becomes optional. Some cases in which field theory might prove essential, rather than merely convenient, in a many-body problem are these: problems in which photons are emitted or absorbed, or in which the use of an instantaneous Coulomb interaction is inadequate (here only the electromagnetic field would need to be quantized, and indeed, real emission and absorption processes can often be treated semiclassically); description of a high-energy plasma in which pair production becomes important; treatment of the nuclear force in nuclei in terms of virtual meson exchange [1] rather than an instantaneous two-body interaction.

There are several ways of handling many-particle systems using the quantum field theory approach [2–14]. We shall here confine our attention to the perturbation expansion, that is, the expansion of quantities of interest in powers of the interaction strength. While many of the most important treatments, such as the Landau quasiparticle theory [15] the BCS–Bogoliubov theory of superconductivity [16, 17] and many others [18–25], avoid the perturbation expansion, it is felt that this approach provides a framework within which most other treatments may be understood, especially those which arise as natural extensions of perturbation theory. Such an extension might arise, for example, in the use of an integral equation to describe the sum of a physically important infinite subset of the terms in the expansion.

There are several ways of setting up the perturbation expansion: some [26–33] deal directly with properties of the system as a whole, such as total energy, or thermodynamic functions, while others [34–38] deal with certain microscopic correlation functions, from which the macroscopic properties also can be obtained. We shall work with one of the latter methods of approach and study the properties of, and methods of calculating the functions known as propagators, which describe the propagation, interaction, damping, and space and time correlations of one or more single-particle (or particle-like) excitations of the system. This can be done for the ground state of the system and, with some modifications but in essentially the same form, for the system at a finite temperature. While it is always possible and sometimes necessary to regard the ground state problem as merely a special case of the finite-temperature problem, we shall here treat the ground state problem first and in greater detail, as being the simpler and the more illuminating.

The boson system with a condensed phase, that is, with one single-particle state occupied by a number of particles proportional to the volume, requires special treatment [39–41]. It turns out that this can be reduced to an equivalent problem in which the macroscopically occupied state plays no dynamical rôle, while the number of particles is not conserved. This equivalent problem can be handled in a straightforward way by the methods we shall develop.

2

Second quantization

WE SHALL approach the problem of introducing second quantization from a point midway between what might be regarded as the two standard methods; these two methods, while mathematically equivalent, are conceptually extremely different, and indeed, they approach the fact of particle-wave duality from opposite directions, since one method [42] takes the field as the fundamental concept, and derives the particle behavior as a consequence, while the other [43] takes the particle states as basic, and relegates the field concept to a secondary rôle.

The first of these standard approaches is to start with a classical field, which is a system with an infinite number of degrees of freedom, and to apply the canonical quantization procedure. That is, one starts with the classical Hamiltonian description of the field, with a complete, and in this case infinite, set of canonical variables; one then replaces each variable by a quantum mechanical operator, with each such operator having a commutator $i\hbar$ with its canonically conjugate operator, and commuting with all the others. The quantum mechanical Hamiltonian is taken as being the same function (or functional) of the operator fields as the classical Hamiltonian is of the corresponding classical field variables. There is often some arbitrariness about the order of noncommuting factors in the operator Hamiltonian; this poses only formal problems in general.

The states of excitation of the system are then found to behave like particle states, as we have noted, and we can identify the quanta with the particles of nature, provided the properties of the original classical field, which are tied to the form of the classical Lagrangian, are so chosen that the quanta will have the desired particle properties. To make these quanta behave like ordinary nonrelativistic particles, in particular, it is necessary to take a classical field equation identical in form to the single-particle Schrödinger equation. The fact that this equation arises originally as the result of quantization of the single-particle system is the origin of the somewhat misleading phrase "second quantization", which is thus not to be taken as referring to a further quantization of a system already quantized, but rather to the formal use of the Schrödinger wave equation (or the Dirac equation, etc.) as a classical field equation.

The other standard approach makes use of what is called the "occupation number representation", and permits one to avoid the field concept altogether. The states are characterized by a specification of how many particles there are in each of a complete set of single-particle states*, and one simply *defines* abstract operators which have the property of increasing or decreasing the number of particles present in a given state. That is, a creation operator is a transformation which takes every N-particle state into an $(N+1)$-particle state, while an annihilation operator is a transformation which takes every N-particle state into an $(N-1)$-particle state. No physical production or absorption process is implied by the use of these operators, which are mathematical in character; nonetheless, they permit the description of such production or absorption processes when they do occur.

The creation and annihilation operators arise in a natural way in the field theory approach also, being equivalent or directly related to the spatially localized field operators with which one starts. They appear most naturally, for the noninteracting field, in connection with the field's normal modes of oscillation, which behave as we have noted, like harmonic oscillators, and are quantized in the same way. Since the excitation of one of these normal modes to a higher state corresponds to the addition of a particle when the quanta are regarded as particles, we see that the raising and lowering operators for this normal mode, regarded as a harmonic oscillator, are just the particle creation and annihilation operators. *We approach the problem, then, at this point,* namely, by relating each single-particle state formally to a harmonic oscillator, and using the operators appropriate for describing these oscillators as the basic operators of our theory. As we shall see, this description as it stands will serve only for bosons, and we shall introduce a modi-

* We make this arbitrary distinction between the phrases "single-particle state" and "one-particle state": the latter refers to any state of the system in which just one particle is present, while the former refers to any one of a number of possible states for a single particle. For example, one might have a one-particle state in which the particle was in a given single-particle state, such as a momentum eigenstate, or one might have an N-particle state in which all the particles were in the same single-particle state, or in N different single-particle states.

fied system, somewhat arbitrarily called the Dirac oscillator, to permit the description of fermion states.

2.1 *The harmonic oscillator—Boson systems*

We start with a description of the harmonic oscillator in terms of the raising and lowering operators $a^\dagger$ and a [44]. We do not attempt to relate our picture closely to the conventional harmonic oscillator, since our interest is confined to the property of having equally spaced energy levels. Indeed, the operator of primary usefulness to us, apart from a and $a^\dagger$ themselves, is the operator $\mathbf{n}$ whose eigenvalues are the integers 0, 1, 2, ..., labeling the various excited states of the oscillator. These eigenvalues, in the particle picture, tell the number of particles present in the corresponding single-particle state.

Our basic assumptions are the commutation relation

$$[a, a^\dagger] = 1, \tag{2.1}$$

and the form of $\mathbf{n}$

$$\mathbf{n} = a^\dagger a. \tag{2.2}$$

The operator $\mathbf{n}$ is clearly hermitian, and its eigenvalues n and normalized eigenvectors ψ_n satisfy

$$\mathbf{n}\psi_n = n\psi_n. \tag{2.3}$$

The eigenvalue n is nonnegative, as can be seen by expressing it as the expectation value of $\mathbf{n}$ in one of its eigenstates:

$$n = \psi_n^\dagger a^\dagger a\, \psi_n$$

$$= |a\psi_n|^2 \tag{2.4}$$

$$\geqslant 0. \tag{2.5}$$

Clearly

$$n = 0 \tag{2.6}$$

if and only if

$$a\psi_n = 0. \tag{2.7}$$

Because of the commutation relation (2.1), the product $aa^\dagger$ can be written in terms of $\mathbf{n}$ thus:

$$aa^\dagger = \mathbf{n} + 1, \tag{2.8}$$

so that $aa^\dagger$ has the eigenvalue $n + 1$ in the state ψ_n. Its eigenvalues are all greater than or equal to one, therefore, and so we see that $a^\dagger\psi_n$ cannot be zero, since

$$|a^\dagger\psi_n|^2 = \psi_n^\dagger aa^\dagger\psi_n$$

$$= n + 1 \tag{2.9}$$

$$\neq 0. \tag{2.10}$$

Now the operator a has the property of lowering the value of n by unity, as we shall now show, and from this the eigenvalues of $\mathbf{n}$ can be deduced. We notice that using Eqs. (2.2, 8) we can write the expression $aa^\dagger a$ in terms of $\mathbf{n}$ in two different ways:

$$aa^\dagger a = (\mathbf{n} + 1)\, a;$$

$$aa^\dagger a = a\mathbf{n}. \tag{2.11}$$

In consequence, we find, ψ_n being an eigenvector, that

$$(\mathbf{n} + 1)\, a\psi_n = a\mathbf{n}\psi_n$$

$$= na\psi_n, \tag{2.12}$$

or

$$\mathbf{n}a\psi_n = (n - 1)\, a\psi_n. \tag{2.13}$$

Thus either $a\psi_n$ is zero and, by Eqs. (2.6, 7), $n = 0$, or else $a\psi_n$ is an eigenvector of $\mathbf{n}$ with eigenvalue $n - 1$. In similar fashion, if n is neither 0 or 1, $a^2\psi_n$ is an eigenvector and the corresponding eigenvalue is $n - 2$. Continuing the process, we see that unless n is one of the integers 0, 1, 2, ..., we can generate negative eigenvalues of $\mathbf{n}$ by a sufficient number of applications of the lowering operator a. This is contrary to the previous result (Eq. 2.5) that the eigenvalues of $\mathbf{n}$ are nonnegative, so we can conclude that the eigenvalues n of $\mathbf{n}$ are all nonnegative integers.

By an argument similar to Eqs. (2.11–13), we have

$$\mathbf{n}a^\dagger\psi_n = a^\dagger aa^\dagger\psi_n$$

$$= a^\dagger\,(\mathbf{n} + 1)\,\psi_n$$

$$= (n + 1)\, a^\dagger\psi_n, \tag{2.14}$$

so that $a^\dagger\psi_n$, which cannot be zero (Eq. 2.10), is certainly an eigenvector of $\mathbf{n}$ with eigenvalue $n + 1$. Thus every nonnegative integer is an eigenvalue of $\mathbf{n}$, since from any one integral eigenvalue n we can generate all others by successive applications of a or $a^\dagger$. Any degeneracy of these levels would correspond to additional degrees of freedom, dynamically independent of **n**, and thus be irelevant to the present discussion.

The normalization of the eigenvectors $a\psi_n$ and $a^\dagger\psi_n$ can be obtained from Eqs. (2.4, 9), and with appropriate choice of phases we can write

$$a\psi_n = n^{1/2}\psi_{n-1}; \tag{2.15}$$

$$a^\dagger\psi_{n-1} = n^{1/2}\psi_n. \tag{2.16}$$

These are for $n = 1, 2, \ldots$; Eq. (2.15) can be taken as valid for $n = 0$ if the right side is taken to be zero regardless of the nonexistence of ψ_{-1}. The eigenvector ψ_n can of course be obtained from ψ_0 by successive applications of $a^\dagger$, so that

$$\psi_n = (n!)^{-1/2}(a^\dagger)^n\psi_0. \tag{2.17}$$

The Hamiltonian for the harmonic oscillator will have the proper spectrum of eigenvalues if we take it as simply

$$\begin{aligned} H &= \varepsilon \mathbf{n} \\ &= \varepsilon a^\dagger a, \end{aligned} \tag{2.18}$$

where ε is the desired spacing of the energy levels, equal to $h\nu_c$ if ν_c is the classical frequency of oscillation. An additive constant $\frac{1}{2}\varepsilon$, which usually appears on quantizing the classical oscillator, but which is not required by the correspondence principle, is here left out. The eigenstates ψ_n of **n** are seen to be the energy eigenstates also:

$$H\psi_n = n\varepsilon\psi_n. \tag{2.19}$$

Our procedure now is to associate one of these oscillators with each possible single-particle state—more precisely, with each of a complete orthonormal set of single-particle states of definite energy (notice the "second quantization" effect again: one needs to have quantized the one-particle system to be able to talk about single-particle states). We shall use a label k to specify these single-particle states; for now they may be any complete set, but we shall later take them to be momentum eigenstates, with k representing the wave vector. We assume they are discrete, although the extension to continuous states is mathematically straightforward. The excitation of the kth oscillator to its n_k^{th} excited state corresponds to the presence of n_k particles in that single-particle state. The operator $\mathbf{n}_k$ whose eigenvalues are n_k is therefore called the "occupation number operator" for that single-particle state.

The different normal modes of the field are dynamically independent, and we express this fact by allowing the operators a_k and $a_k^\dagger$ associated with different values of k to commute. This, together with the assumption (2.1), can be written:

$$[a_k, a_{k'}] = 0; \tag{2.20}$$

$$[a_k, a_{k'}^\dagger] = \delta(k, k') \tag{2.21}$$

where $\delta(x, y)$ represents the Kronecker delta function, equal to one when $x = y$, and to zero when $x \neq y$. The state in which every normal mode is unexcited corresponds to the vacuum state—no particles present in any single-particle state—and is denoted by the vector Φ_0, normalized to unity. We do not need to give an explicit representation of this vector, since it is only the relations between vectors, and the scalar products of vectors, that enter the theory. The vacuum has the property (2.6, 7) with respect to every single-particle state k:

$$a_k\Phi_0 = 0, \quad \text{all } k. \tag{2.22}$$

The states in which the occupation numbers of all the single-particle states (or excitation numbers of all the normal modes) are specified form a complete set of states of the system, and form a convenient orthonormal basis for the description of any arbitrary state. We can use the symbol $\{n\}$ to denote the set of numbers n_k, and can express the state vector $\Phi_{\{n\}}$ in terms of the vacuum state vector Φ_0 by the use of creation operators:

$$\Phi_{\{n\}} = \prod_k [(n_k!)^{-1/2}(a_k^\dagger)^{n_k}]\,\Phi_0. \tag{2.23}$$

The most general possible state would be a linear combination of these basis vectors, thus:

$$\Psi = \sum_{\{n\}} C_{\{n\}}\Phi_{\{n\}}, \tag{2.24}$$

where there is a coefficient $C_{\{n\}}$ for each possible set $\{n\}$ of occupation numbers. Note that states with different values of the total particle number $\Sigma\, n_k$ can be combined in this way, if one should wish to do so. This is necessary in a relativistic theory, and is often a convenient artifice even in a nonrelativistic theory, such as, for example the BSC–Bogoliubov theory of superconductivity [16, 17].

A state in which one particle is present in the single-particle state k would be represented by the vector

$$\Phi_k = a_k^\dagger\Phi_0, \tag{2.25}$$

while a two-particle state vector might have the form

$$\Phi_{k,k'} = a_k^\dagger a_{k'}^\dagger\Phi_0. \tag{2.26}$$

The ground state of the N-particle system (without interaction) would correspond to the state vector

$$\Phi_{N,0} = (N!)^{-1/2}(a_0^{\dagger})^N\Phi_0, \tag{2.27}$$

where $k = 0$ refers to the single-particle state of lowest energy.

On looking back at what we have done so far, we see that we have arrived at a suitable description of a system of bosons. The particles are certainly indistinguishable since our description of a state specifies only the occupation numbers and does not, indeed cannot, identify individual particles. As a consequence of the harmonic oscillator character of the normal modes of the field, we find that the occupation number of a single-particle state may be any nonnegative integer; there is no exclusion principle. The two-particle state (2.26) is symmetric under interchange of k and k', and this symmetry extends to states of any number of particles. In short, the quanta of a classical field inevitably satisfy Bose-Einstein (BE) statistics.

2.2 The Dirac oscillator—Fermion systems

To describe fermions, we have to abandon the normal mode concept associated with the classical field and work only with the particle picture. (A quantum operator field can be introduced, but it has no classical limit in the sense of an observable classical field.) As before, we associate raising and lowering operators $a^{\dagger}$ and a with each single-particle state, but they do not have to describe a harmonic oscillator. Our basic unit system for the Fermi-Dirac case, corresponding to the harmonic oscillator in the Bose-Einstein case, might be called a Dirac oscillator [44], and the basic operators a, $a^{\dagger}$ and $\mathbf{n}$ are related now by

$$[a, a^{\dagger}]_+ \equiv aa^{\dagger} + a^{\dagger}a = 1; \tag{2.28}$$

$$[a, a]_+ \equiv 2aa = 0; \tag{2.29}$$

$$\mathbf{n} = a^{\dagger}a. \tag{2.30}$$

These are to be compared with Eqs. (2.1, 2), where the new equation (2.29) replaces the obvious statement in the Bose–Einstein case that the operator a commutes with itself. We again denote the eigenvalues of $\mathbf{n}$ by n, and the normalized eigenvectors by ψ_n:

$$\mathbf{n}\psi_n = n\psi_n. \tag{2.31}$$

Following Eqs. (2.4, 5), we again find that the eigenvalue n satisfies

$$\begin{aligned} n &= \psi_n^\dagger a^\dagger a \psi_n \\ &= |a\psi_n|^2 \qquad (2.32) \\ &\geqslant 0, \qquad (2.33) \end{aligned}$$

so that n is nonnegative, while

$$n = 0 \qquad (2.34)$$

if and only if

$$a\psi_n = 0. \qquad (2.35)$$

When we use Eqs. (2.28, 30) to express $aa^\dagger$ in terms of $\mathbf{n}$, however, we find that

$$aa^\dagger = 1 - \mathbf{n}, \qquad (2.36)$$

in distinction to Eq. (2.8), so that

$$\begin{aligned} 1 - n &= \psi_n^\dagger a a^\dagger \psi_n \\ &= |a^\dagger \psi_n|^2 \qquad (2.37) \\ &\geqslant 0. \qquad (2.38) \end{aligned}$$

Thus n cannot exceed unity, and

$$n = 1 \qquad (2.39)$$

if and only if

$$a^\dagger \psi_n = 0. \qquad (2.40)$$

We can now use Eqs. (2.29) together with the eigenvalue equation (2.31) to deduce the eigenvalues of $\mathbf{n}$:

$$\begin{aligned} 0 &= a^\dagger a^\dagger a \psi_n \\ &= a^\dagger \mathbf{n} \psi_n \\ &= n a^\dagger \psi_n, \qquad (2.41) \end{aligned}$$

so that for each ψ_n either

$$n = 0, \qquad (2.42)$$

in which case (Eq. 2.34, 35)

$$a\psi_n = 0, \qquad (2.43)$$

or else

$$a^\dagger \psi_n = 0, \qquad (2.44)$$

in which case (Eqs. 2.39, 40)

$$n = 1. \tag{2.45}$$

Thus the only possible eigenvalues of $\mathbf{n}$ are 0 and 1, and if either is an eigenvalue, the other can be obtained by application of a or $a^\dagger$:

$$\begin{aligned} \mathbf{n} a^\dagger \psi_0 &= (1 - a a^\dagger)\, a^\dagger \psi_0 \\ &= a^\dagger \psi_0 , \end{aligned} \tag{2.46}$$

while

$$\begin{aligned} \mathbf{n} a \psi_1 &= a^\dagger a a \psi_1 \\ &= 0. \end{aligned} \tag{2.47}$$

In short, we have two states (excepting possible degeneracy, which would again be associated with additional independent degrees of freedom), satisfying, with appropriate choice of phases,

$$a\psi_0 = 0; \tag{2.48}$$

$$a\psi_1 = \psi_0; \tag{2.49}$$

$$a^\dagger\psi_0 = \psi_1; \tag{2.50}$$

$$a^\dagger\psi_1 = 0. \tag{2.51}$$

A natural Hamiltonian to choose for the Dirac oscillator is

$$\begin{aligned} H &= \varepsilon \mathbf{n} \\ &= \varepsilon a^\dagger a , \end{aligned} \tag{2.52}$$

in analogy to that for the harmonic oscillator (Eq. 2.18); the energy eigenvalues are then simply 0 and ε.* However, since $\mathbf{n}$ is bounded above as well

---

* The oscillator character of the Dirac oscillator, while extraneous to our discussion, can be brought out by defining hermitian "position" and "momentum" operators

$$x = \hbar\,(2m\varepsilon)^{-1/2}\,(a + a^\dagger);$$

$$p = -i\,(\tfrac{1}{2}m\varepsilon)^{-1/2}\,(a - a^\dagger);$$

(the same relations as for the harmonic oscillator), with m an arbitrary parameter with the dimension of mass. These are found to have the following properties:

$$xp + px = 0;$$

$$x^2 = \hbar^2\,(2m\varepsilon)^{-1}; \quad p^2 = \tfrac{1}{2}m\varepsilon;$$

so that x and p each have just two discrete eigenvalues. If we take for the Hamiltonian $H = \varepsilon a^\dagger a$, as in Eq. (2.52), then the expectation value of x undergoes a sinusoidal variation with frequency ε/h, and amplitude dependent on initial conditions, but no greater, interestingly, than $\hbar\,(2m\varepsilon)^{-1/2}$. The expectation values of p and x are related by the expressions $d\langle x\rangle/dt = \langle p\rangle/m$, and $d\langle p\rangle/dt = -m\,(\varepsilon/\hbar)^2\,\langle x\rangle$, which are identical to the corresponding expressions for the harmonic oscillator.

as below, it is possible, and often useful, to reverse the order of the levels and take

$$H = \varepsilon(1 - \mathbf{n})$$
$$= \varepsilon a a^\dagger; \tag{2.53}$$

this reverses the roles of a and $a^\dagger$, which is perfectly possible since all of their relations (Eqs. 2.28, 48–51) are symmetrical between them.

We now associate one Dirac oscillator with each single-particle state k. The excitation of any of these to its $n_k = 1$ state corresponds to the presence of a particle in that single-particle state. Since $n_k = 2$ does not occur as an eigenvalue of the occupation number operator $\mathbf{n}_k$, we see that the exclusion principle is satisfied. We can satisfy the requirement of antisymmetry by making (in fact, *only* by making) the operators a_k and $a_k^\dagger$ anticommute for different values of k:

$$[a_k, a_{k'}]_+ = 0; \tag{2.54}$$

$$[a_k, a_{k'}^\dagger]_+ = \delta(k, k'), \tag{2.55}$$

These include Eqs. (2.28, 29) in a natural way, for the case $k = k'$. The vacuum state Φ_0 still satisfies an equation identical in form to Eq. (2.22):

$$a_k \Phi_0 = 0 \quad (\text{all } k), \tag{2.56}$$

and the one-particle state is again given by $a_k^\dagger \Phi_0^\dagger$. The two-particle state analogous to (2.26) is now seen to be antisymmetric:

$$a_k^\dagger a_{k'}^\dagger \Phi_0 = -a_{k'}^\dagger a_k^\dagger \Phi_0, \tag{2.57}$$

and the general state, of any number of particles, is also completely antisymmetric. Again, an arbitrary state vector can be expressed in terms of the complete set of vectors (2.23), where now the occupation numbers are restricted to the values 0 and 1, and the sign of $\Phi_{\{n\}}$ is dependent on the order in which the factors $a_k^\dagger$ are written. It is clear, then, that this provides a description of particles obeying Fermi–Dirac (FD) statistics.

The ground state of a system of N noninteracting fermions is described by the state vector

$$\Phi_{N,0} = \prod_{k=1}^{N} a_k^\dagger \Phi_0, \tag{2.58}$$

where the product notation indicates (vaguely) that k is to run through the

N single-particle states of lowest energy. It has the properties, similar to Eq. (2.56), that

$$a_k \Phi_{N,0} = 0 \quad (k > N); \tag{2.59}$$

$$a_k^\dagger \Phi_{N,0} = 0 \quad (k \leqslant N). \tag{2.60}$$

It has been noted in connection with Eq. (2.53) that the relationships of a_k and $a_k^\dagger$ to each other are symmetrical, so that their rôles may be interchanged provided $\mathbf{n}_k$ is replaced by $1 - \mathbf{n}_k$. This means that for $k \leqslant N$, if we wish, we can think of $a_k^\dagger$ as the annihilation operator for "holes", a_k as the hole creation operator, and $\Phi_{N,0}$ as a "vacuum", with neither particles ($k > N$) nor holes ($k \leqslant N$) present. (The symmetry between particles and holes in real systems, for excitations close to the Fermi surface, has been investigated in terms of an actual symmetry operation by Fisher [45].)

It should be noted that while the mathematical descriptions of fermions and bosons differ only slightly—a difference in sign in the relations (2.20, 21, 54, 55) and a restriction on the values n_k, yet the physical differences may be very great, in terms of both practical behavior and fundamental nature. The fact that the BE system can be obtained by quantizing a classical field implies the converse: that there is a classical field limit to the system of bosons which does not exist for a system of fermions.

2.3 *The Hamiltonian*

We have so far said nothing about the dynamics of the many-particle system we are attempting to describe, but have simply provided a complete description of the possible states of the system. This description differs from the Schrödinger wave function picture in that states with different numbers of particles are now described by vectors in the same space, and these vectors can thus be superposed if desired. The vectors corresponding to some definite number N of particles are in exact one-to-one correspondence with the symmetrized (or antisymmetrized) N-particle wave functions of the ordinary Schrödinger theory.

Because of this correspondence it is trivial in principle to construct linear operators for the new description which have the same matrix elements between corresponding state vectors as the familiar operators, such as energy and momentum, in the Schrödinger picture. It is a little more complicated to construct these corresponding operators explicitly, and to show that in most cases they take on a very simple form which is the same for different values of the total particle number N.

The single-particle properties are easy to express in terms of the occupation number operators $\mathbf{n}_k$. If the energy of a single particle in the state k is ε_k (which might be the kinetic energy, for example, in the case that k refers to the momentum), then the sum of the single-particle energies in the state $\Phi_{\{n\}}$ (Eq. 2.23) is

$$E^0_{\{n\}} = \sum_k n_k \varepsilon_k. \tag{2.61}$$

The operator that has these eigenvalues is just

$$\begin{aligned} H_0 &= \sum_k \varepsilon_k \mathbf{n}_k \\ &= \sum_k \varepsilon_k a_k^\dagger a_k. \end{aligned} \tag{2.62}$$

This independent-particle Hamiltonian, which we shall have occasion to use later on as the unperturbed Hamiltonian for a perturbation expansion, is just the sum of oscillator Hamiltonians (Eqs. 2.18, 52), for the separate single-particle states. Similarly, the total number of particles

$$N_{\{n\}} = \sum_k n_k \tag{2.63}$$

can be regarded as the eigenvalue of the total number operator $\mathbf{N}$, given by

$$\begin{aligned} \mathbf{N} &= \sum_k \mathbf{n}_k \\ &= \sum_k a_k^\dagger a_k. \end{aligned} \tag{2.64}$$

If $\hbar k$ is a linear momentum vector, then the total momentum in the state $\Phi_{\{n\}}$ is just

$$P_{\{n\}} = \sum_k \hbar k n_k, \tag{2.65}$$

which is the appropriate eigenvalue of the total momentum operator

$$P = \sum_k \hbar k \mathbf{n}_k. \tag{2.66}$$

This last is not true, of course, if the single particle states k do not correspond to definite values of the linear momentum. Other single-particle properties, such as charge and angular momentum, for example, are fairly straightforward to express in this way.

The interaction part of the Hamiltonian is by no means so obvious, and ts form depends, of course, on the physical interaction being considered.

Two common cases are presented here without derivation. The simple two-body interaction, which in the Schrödinger picture would be

$$\mathscr{H}_1 = \sum_{i<j} v\,(x_i - x_j), \tag{2.67}$$

becomes in our picture the operator

$$H_1 = \tfrac{1}{2} \sum a_1^\dagger a_2^\dagger v\,(k_1, k_2; k_1', k_2')\, a_{2'} a_{1'}, \tag{2.68}$$

where the subscripts 1, 2, 1′, 2′, are short for the labels k_1, k_2, k_1', and k_2', (which still need not be momenta, and in particular will include any spin parameters). The sum is over all values of the four parameters k, and $v\,(k_1, k_2; k_1', k_2')$ is the matrix element of $v\,(x - x')$ between the two-particle states (k_1, k_2) and (k_1', k_2'):

$$v\,(k_1, k_2; k_1', k_2') = \int u_1^*(x_1)\, u_2^*(x_2)\, v\,(x_1 - x_2)\, u_1'\,(x_1)\, u_2'\,(x_2)\, d^3x_1\, d^3x_2\,. \tag{2.69}$$

This need not be symmetrized or antisymmetrized, the necessary symmetry of H_1 being provided by the operators a and $a^\dagger$ (Eqs. 2.20, 54). The extension to spin-dependent or nonlocal interactions is straightforward, and quite generally, if the k's are linear momenta, the matrix element $v_{12,1'2'}$ can be expressed in terms of relative and total momenta thus:

$$v\,(k_1, k_2; k_1', k_2') = (2\pi)^3\, V^{-1}\, \delta(K, K')\, v\,(k, k'; K), \tag{2.70}$$

where

$$\begin{aligned} k &= \tfrac{1}{2}\,(k_1 - k_2), \\ k' &= \tfrac{1}{2}\,(k_1' - k_2'), \\ K &= k_1 + k_2, \\ K' &= k_1' + k_2', \end{aligned} \tag{2.71}$$

and $v\,(k, k'; K)$, which reduces to $v\,(k - k')$ for local potentials, remains finite as the volume V becomes infinite. The significance of the operator H_1 is perhaps best understood in terms of the virtual processes to which it gives rise. To any one term in the sum in Eq. (2.68) there would correspond a virtual process in which two particles are scattered from states k_1', k_2' into states k_1, k_2. This is performed mathematically, when H_1 operates on a state in which two such particles are present, by the annihilation of the two particles in states k_1' and k_2', and the creation of two more in the states k_1 and k_2, with a weighting factor equal to the matrix element $v_{12,\,1'2'}$ for such a transition. When the k's are momenta, the factor $\delta\,(k, k')$ (Eq. 2.70) in the matrix element ensures that the total momentum is the same in the final state as in the initial state. Note that H_1, like H_0, has a form which is independent of the total number of particles N. This is particularly convenient when the limiting properties of very large systems are being studied, or where it is desirable to leave the total particle number indefinite.

Another example of common interest is an interaction in which a particle can be produced or absorbed, as in the interaction of electrons and phonons in a solid [46]. In this case there is no Schrödinger-wave-function analog, and the interaction Hamiltonian takes the general form

$$H_1 = \sum_{k,k',q} a_k^\dagger a_{k'} (v_{kk',q} b_q + v_{k'k,q}{}^* b_q^\dagger), \tag{2.72}$$

where b_q and $b_q^\dagger$ are operators exactly like a_k and $a_k^\dagger$, but describing the annihilation or creation of a second type of particle (e.g., the phonon), which must be a boson to conserve statistics. The particle described by a_k and $a_k^\dagger$ may be either a boson or a fermion (the electron in our example). The states q are to be regarded as a different set from the states k, since they are associated with a different kind of particle. An additional term is needed in H_0 to complete the description of the additional type of particle, of the same form as Eq. (2.62), but with a different single-particle energy spectrum. The electron single-particle states in our example, would best be taken as Bloch waves, presumably, with an appropriate energy spectrum. Again, both k and q may refer to linear momenta, and the interaction matrix elements $v_{kk',q}$ may conserve momentum in the interaction. The electron–electron interaction associated with the virtual emission and reabsorption of a phonon will appear first in second-order perturbation theory if we use this interaction Hamiltonian.

2.4 *Time-dependent operators*

We now consider the time-dependence of our system, starting with the abstract vector form of the Schrödinger equation:

$$i\hbar \, d\Psi(t)/dt = H\Psi(t). \tag{2.73}$$

Here the Hamiltonian H is given by

$$H = H_0 + H_1, \tag{2.74}$$

and the independent-particle part H_0 and the interaction part H_1 are given by Eqs. (2.62) and (2.68) or (2.72). The differential equation (2.73) can be integrated formally, in the case that H does not involve the time explicitly, to give

$$\Psi(t) = e^{-iHt}\Psi(0) \tag{2.75}$$

(we shall suppose that physical quantities have been defined in such a way that action is dimensionless and $\hbar = 1$). Equation (2.75) means simply that if $\psi(0)$ is expressed in the form

$$\Psi(0) = \sum_{\nu} c_{\nu}\Psi_{\nu}, \tag{2.76}$$

where Ψ_{ν} are the eigenstates of H with eigenvalues E_{ν}, then

$$\Psi(t) = \sum_{\nu} c_{\nu} e^{-iE_{\nu}t}\Psi_{\nu}. \tag{2.77}$$

These relations define the operator e^{-iHt} quite satisfactorily, and it is seen to be unitary, i.e.

$$(e^{-iHt})^{-1} = (e^{-iHt})^{\dagger}. \tag{2.78}$$

Now our method of studying the system is roughly simply this: to observe the behavior of the system with time when it is perturbed from the ground state, or from statistical equilibrium, by the addition or subtraction of one, or perhaps two, particles. In the case of the addition to the state Ψ of a particle in the state k at the time t_1, the behavior with time would then be given by

$$\begin{aligned}\Psi_{kt_1}(t) &= e^{-iH(t-t_1)}a_k^{\dagger}\Psi(t_1)\\ &= e^{-iHt}\,[e^{iHt_1}a_k^{\dagger}\,e^{-iHt_1}]\,\Psi(0).\end{aligned} \tag{2.79}$$

The bracketed expression depends only on t_1, and for the case $t = 0$ we have

$$\Psi_{kt_1}(0) = \mathbf{a}_k^{\dagger}(t_1)\,\Psi(0), \tag{2.80}$$

where the bold-face time-dependent operator is defined quite generally (for arbitrary operator A) by

$$\mathbf{A}(t) = e^{iHt}Ae^{-iHt}. \tag{2.81}$$

Such an operator $\mathbf{A}(t)$, which is related to A by a time-dependent unitary transformation, is said to be in the "Heisenberg representation", while the original operator A is in the "Schrödinger representation". The Heisenberg operators $\mathbf{a}_k^{\dagger}(t)$ and $\mathbf{a}_k(t)$ play an important rôle in the subsequent development. Any Heisenberg operator $\mathbf{A}(t)$ satisfies the Heisenberg equations of motion, obtained by differentiating Eq. (2.81)

$$i\,d\mathbf{A}(t)/dt = [\mathbf{A}(t), H]. \tag{2.82}$$

The Hamiltonian H itself is unaltered by the transformations (2.81), since it commutes with the unitary operator e^{-iHt}; from this, or from Eq. (2.82) (with $A = H$), it follows that H is still time-independent in the Heisenberg representation. The two parts H_0 and H_1, however, would each be time-dependent in the Heisenberg representation, even though their sum is not.

The operator $\mathbf{a}_k(t)$, in particular, satisfies an equation of motion which can be obtained from Eq. (2.82), the commutation relations (2.20, 21) or (2.54, 55), and the specific form of the Hamiltonian. For the case of a particle–particle interaction (Eq. 2.68), this is

$$i\dot{\mathbf{a}}_1 = \varepsilon_1 \mathbf{a}_1 + \sum_{1'22'} v(k_1, k_2; k_1', k_2')\, \mathbf{a}_2^\dagger \mathbf{a}_2{}' \mathbf{a}_1{}', \tag{2.83}$$

where again the subscripts 1, 1′, etc., are short for the parameters k_1, k_1', etc. It is useful, as we shall see later, to note that this permits the Hamiltonian (2.62, 68) to be written in a concise way in terms of the Heisenberg operators:

$$H = \tfrac{1}{2} \sum_k \mathbf{a}_k^\dagger (i\dot{\mathbf{a}}_k + \varepsilon_k \mathbf{a}_k). \tag{2.84}$$

Similar, but often more complicated, relations hold for other types of interaction, such that of as Eq. (2.72).

We shall have occasion also to make a transformation similar to Eq. (2.81) but with H replaced by H_0. This gives rise to what is called the "interaction representation"; for any Schrödinger operator A, the corresponding interaction representation operator $A(t)$ is given by

$$A(t) = e^{iH_0 t} A e^{-iH_0 t}, \tag{2.85}$$

and satisfies

$$i\, dA(t)/dt = [A(t), H_0]. \tag{2.86}$$

The independent particle Hamiltonian H_0 is unaltered by the transformation (2.85), and is time-independent, while the total Hamiltonian $H(t)$ is now time-dependent. The Heisenberg representation coincides with the interaction representation, of course, in the case of noninteracting particles, for which $H = H_0$.

The form of the annihilation operator $a_k(t)$ in the interaction representation is easily obtained from the differential equation (2.86), which becomes (using interaction representation operators in H_0)

$$\begin{aligned} i\, da_k(t)/dt &= [a_k(t), \sum_{k'} \varepsilon_{k'} a_{k'}^\dagger(t)\, a_{k'}(t)] \\ &= \varepsilon_k\, [a_k(t), a_k^\dagger(t)\, a_k(t)] \\ &= \varepsilon_k a_k(t). \end{aligned} \tag{2.87}$$

Here we have used the fact that the commutation relations (2.20, 21, 54, 55) are unaltered by the unitary transformation; the result is the same for both FD and BE statistics. This differential equation is trivially soluble, using the boundary condition at $t = 0$ (from Eq. 2.85)

$$a_k(0) = a_k; \tag{2.88}$$

the solution is

$$a_k(t) = e^{-i\varepsilon_k t} a_k. \tag{2.89}$$

To see this, note that $d\,[e^{i\varepsilon_k t} a_k(t)]/dt = 0$, so that $e^{i\varepsilon_k t} a_k(t)$ is an operator constant in time.

3

Propagators—General properties

WE SHALL use various functions, all closely related, which we shall indiscriminately call "propagators",* and which, as has been mentioned, describe the behavior of the system when perturbed by the addition or subtraction of one or more particles. We shall confine our attention to the case of just one particle; the generalization of the perturbation expansion method to the case of several particles is straightforward. Our propagators are defined basically as expectation values of products of annihilation and creation operators and may be calculated for the true system or for the unperturbed system.

Roughly three cases present themselves: the ground state, the finite temperature case in which the particle number is held fixed, and the finite temperature case in which there is no restriction on the particle numbers. For the case of the ground state the same techniques can be used whether or not the total particle number is conserved; if it is conserved, our expectation values are taken with respect to the N-body ground state. At a finite temperature we use statistical expectation values, taken with respect to either the canonical or the grand canonical ensemble, the latter giving an average without restriction on the particle number, in the case that the particle number is conserved. If the particle number is not conserved, that is, if absorption and production processes are possible, then the canonical ensemble is prescribed, but gives an average without particle number restriction. When we pass to the limit of no interactions in this case we must retain the feature of no particle-number restriction, even though particle number is now con-

* These are often referred to as 'Green's functions", since some of the propagators for non-interacting particles are in fact Green's functions for the partial differential equation, in our case the Schrödinger equation, which describes the original field.

served, so that we must deal with the grand canonical ensemble for the unperturbed system. Now the perturbation method which we shall develop for systems at a finite temperature is applicable only when statistical averages are taken without restriction on the particle number; we find it necessary, then, for several reasons, to treat together the canonical ensemble for the particle nonconserving case and the regular grand canonical ensemble for the particle conserving case. To avoid cumbersome verbiage, therefore, we shall refer to both of these by the ad hoc term "grand ensemble", and reserve the phrase "canonical ensemble" for the special case that the particle number is held fixed.

Two important examples of particle nonconserving systems may be mentioned: one is the electron–phonon system (Eq. 2.72), where the electrons are conserved, and therefore may be treated in the grand canonical ensemble, while the phonons are not conserved; the entire system may be described, then, by a grand ensemble. The other case arises from the problem of the many-boson system with a condensed phase, such as liquid helium below the lambda point; here there is one single-particle state, the zero-momentum state, which is macroscopically occupied—that is, its mean occupation number is proportional to the volume in the large-volume limit. In this case (see the introduction to Sec. 5.7) we have to start with averages in the canonical ensemble [39, 41]; this makes it difficult to use our perturbation expansion technique, and we resort to a theorem, discussed in Secs. 4.6 and 5.7, which reduces the problem to an equivalent artificial problem, more easily treated. In the equivalent artificial system the zero-momentum state (the one that is macroscopically occupied) plays no dynamical rôle; its occupation number is treated as a fixed *c*-number, and the scattering of particles into or out of the zero-momentum state is treated as absorption or production of non-zero-momentum particles. The particle number (number of non-zero-momentum particles) is not conserved in this equivalent problem, and so it too can be treated by means of the grand ensemble.

In all of the three basic cases we denote the expectation value of an arbitrary operator A by $\langle A \rangle$:

$$\langle A \rangle = \Psi^{\dagger}_{N,0}\, A \Psi_{N,0} \quad \text{(ground state)}, \tag{3.1}$$

$$= \frac{\mathrm{Tr}\,[e^{-\beta H} A]_N}{\mathrm{Tr}\,[e^{-\beta H}]_N} \quad \text{(canonical ensemble)}, \tag{3.2}$$

$$= \frac{\mathrm{Tr}\,[e^{-\beta(H-\mu \mathrm{N})} A]}{\mathrm{Tr}\,[e^{--\beta(H-\mu \mathrm{N})}]} \quad \text{(grand ensemble)}. \tag{3.3}$$

Here β is the inverse temperature,

$$\beta = (kT)^{-1}; \tag{3.4}$$

in Eq. (3.2) the trace is over states of given N only, and the denominator is the partition function

$$Q = \mathrm{Tr}\,[e^{-\beta H}]_N. \tag{3.5}$$

We shall restrict most of our discussion to the grand ensemble, noting however that much of it will apply with only minor changes to the canonical ensemble. In Eq. (3.3) the trace is over all states, with arbitrary N, and the denominator is the grand partition function

$$\mathscr{Q} = \mathrm{Tr}\,[e^{-\beta \bar{H}}], \tag{3.6}$$

where

$$\bar{H} = H - \mu \mathbf{N}. \tag{3.7}$$

The parameter μ is the chemical potential, and is chosen to give the desired average number of particles $\mathscr{N}$:

$$\mathscr{N} = \langle \mathbf{N} \rangle; \tag{3.8}$$

alternatively, $\mathscr{N}$ may be thought of as a function of μ and β, as well as the volume and any other parameters involved in $\bar{H}$. If the particle number is not conserved, μ must be set equal to zero, since we are dealing with the canonical ensemble in that case.

Eqs. (3.3, 6) can also be written in terms of the eigenstates and eigenvalues of $\bar{H}$:

$$\bar{H}\Psi_\nu = \bar{E}_\nu \Psi_\nu = (E_\nu - \mu N_\nu)\Psi_\nu; \tag{3.9}$$

$$\langle A \rangle = \mathscr{Q}^{-1} \sum_\nu e^{-\beta \bar{E}_\nu} \Psi_\nu^\dagger A \Psi_\nu. \tag{3.10}$$

Here the case $A = 1$ can be used to describe $\mathscr{Q}$ itself. In the zero-temperature limit β becomes infinite, so that the term in the sum for which $\bar{E}_\nu$ is minimum dominates, and $\langle A \rangle$ becomes the ground state expectation value of A for some particle number N determined by the value of μ. It is therefore reasonable to use the same symbol to describe the statistical expectation value and the ground state expectation value. In this zero-temperature limit, μ becomes the separation energy of a single particle, which for a system with saturating forces is the same as the average energy per particle [47], and for free particles is just the Fermi energy.

3.1 The basic propagators $G^{\pm}(k, t - t_0)$

To describe the propagation of a single-particle excitation, we shall take the scalar product of two state vectors of the type (2.79), with a single particle added (or removed) at the times t and t_0, say: the scalar product is then the probability amplitude for the two states to coincide—that is, roughly, for the single-particle excitation to exist, and to retain its single-particle character from the time t_0 to the time t. Our basic propagation function for a state ν is then (see Eqs. (2.79, 80))

$$\Psi^{\dagger}_{kt}(t')\,\Psi_{k_0 t_0}(t') = [\Psi^{\dagger}_{\nu}(t)\, a_k\, e^{iH(t'-t)}]\,[e^{-iH(t'-t_0)} a^{\dagger}_{k_0} \Psi_{\nu}(t_0)]$$

$$= \Psi^{\dagger}_{\nu} \mathbf{a}_k(t)\, \mathbf{a}^{\dagger}_{k_0}(t_0)\, \Psi_{\nu} \qquad (3.11)$$

(note that the dependence on t' diappears), and the propagator is defined as the statistical average of this (Eqs. 3.3, 10) or else its value for the ground state (Eq. 3.1):

$$G^{+}(k, t; k_0, t_0) = \langle \mathbf{a}_k(t)\, \mathbf{a}^{\dagger}_{k_0}(t_0) \rangle. \qquad (3.12)$$

The propagation of a particle deficiency is described by the related propagator

$$G^{-}(k, t; k_0, t_0) = \langle \mathbf{a}^{\dagger}_{k_0}(t_0)\, \mathbf{a}_k(t) \rangle, \qquad (3.13)$$

where the choice of arguments is such that the excitation is to be regarded as propagating from the time t to the time t_0. These two functions also depend on the possibility of adding or subtracting the extra particle in the first place; that is, G^- will vanish if the state k is vacant at the time t, while G^+ will vanish if, for example, the particles are fermions and the state k_0 is occupied at the time t_0.

We now inspect the form these functions take for non interacting particles (omitting the case of the canonical ensemble, for which the results are more complicated, and not needed for our development). This is equivalent to using the interaction representation operators (Eqs. 2.85, 89) instead of the Heisenberg operators in Eqs. (3.12, 13), and using

$$\bar{H}_0 = H_0 - \mu \mathbf{N} \qquad (3.14)$$

to calculate the statistical expectation values; we denote such an expectation value, for our arbitrary operator A, by

$$\langle A \rangle_0 = \mathcal{Q}_0^{-1}\, \mathrm{Tr}\, [e^{-\beta \bar{H}_0} A]. \qquad (3.15)$$

We get

$$G_0^+(k, t; k_0, t_0) = \langle a_k(t)\, a_{k_0}^\dagger(t_0)\rangle_0 \tag{3.16}$$

$$= \exp\left[-i\,(\varepsilon_k t - \varepsilon_{k_0} t_0)\right] \langle a_k a_{k_0}^\dagger\rangle_0$$

$$= e^{-i\varepsilon_k(t-t_0)}\langle a_k a_k^\dagger\rangle_0\, \delta\,(k, k_0). \tag{3.17}$$

Similarly,

$$G_0^-(k, t; k_0, t_0) = e^{-i\varepsilon_k(t-t_0)}\langle a_k^\dagger a_k\rangle_0\, \delta\,(k, k_0). \tag{3.18}$$

That k must equal k_0 follows from the fact that we are taking expectation values of $a_k a_{k_0}^\dagger$ or $a_{k_0}^\dagger a_k$ in states (Eq. 2.23) in which the single particle states have definite occupation numbers, and if a particle is added to state k_0, for example, it has to be taken away again from state k_0 in order to get back to the original occupation number.

Because of the dynamical independence of the states k in the absence of particle–particle interactions, the expectation value of the occupation number operator $\mathbf{n}_k = a_k^\dagger a_k$ is just the appropriate statistical weighting factor for the ideal gas:

$$\langle \mathbf{n}_k \rangle_0 = \frac{\sum_n n\, e^{-\beta n(\varepsilon k-\mu)}}{\sum_n e^{-\beta n(\varepsilon_k-\mu)}}$$

$$= f^-(\bar{\varepsilon}_k) \tag{3.19}$$

$$\equiv (e^{\beta\bar{\varepsilon}_k} + \sigma)^{-1}. \tag{3.20}$$

The sums over n are from 0 to 1 for Fermi-Dirac statistics, and from 0 to ∞ for Bose–Einstein statistics. We have introduced the shorthand notation

$$\bar{\varepsilon}_k = \varepsilon_k - \mu, \tag{3.21}$$

in analogy to $\bar{H}$ and $\bar{E}_v$, and have let

$$\begin{aligned} \sigma &= +1 \quad \text{(FD statistics)} \\ &= -1 \quad \text{(BE statistics).} \end{aligned} \tag{3.22}$$

We thus find that $G_0^-(k, t; k_0, t_0)$ is given by

$$G_0^-(k, t; k_0, t_0) = \delta\,(k, k_0)\, e^{-i\varepsilon_k(t-t_0)} f^-(\bar{\varepsilon}_k), \tag{3.23}$$

and, since Eqs. (2.8, 36) can be written

$$a_k a_k^\dagger = 1 - \sigma \mathbf{n}_k, \tag{3.24}$$

we get

$$G_0^+(k, t; k_0, t_0) = \delta\,(k, k_0)\, e^{-i\varepsilon_k(t-t_0)} f^+(\bar{\varepsilon}_k), \tag{3.25}$$

where

$$f^{+}(x) = 1 - \sigma f^{-}(x) \tag{3.26}$$

$$= (1 + \sigma\, e^{-\beta x})^{-1} \tag{3.27}$$

$$= e^{\beta x} f^{-}(x). \tag{3.28}$$

We see in this case that the excitation propagates in time in a trivial way in the absence of interaction, and that the probability of being able to form the excitation in the first place is reflected in the statistical factor $f^{\pm}(\bar{\varepsilon}_k)$. The function $f^{-}(\varepsilon_k)$ becomes very small for poorly occupied states, while $f^{+}(\bar{\varepsilon}_k)$ gets small, for fermions, for nearly fully occupied states. For analysis of ground state properties we use the zero-temperature limit ($\beta \to \infty$) of these functions; in this limit both $f^{+}(x)$ and $f^{-}(x)$ become step functions:

$$\begin{aligned} f^{+}(x) &= \theta(x), \\ f^{-}(x) &= \sigma\theta(-x) \quad (\beta \to \infty), \end{aligned} \tag{3.29}$$

where $\theta(x)$ is the unit function:

$$\begin{aligned} \theta(x) &= 1 \quad (x > 0) \\ &= 0 \quad (x < 0). \end{aligned} \tag{3.30}$$

(We do not define $\theta(x)$ for zero argument, but rather require that an appropriate limit be specified in particular cases.)

The circumstance that $k = k_0$ in these free-particle functions, as has been mentioned, reflects the lack of interaction, and not, here, any conservation property. In the absence of such a conservation law, the functions $G^{\pm}(k, t; k_0, t_0)$ will not be diagonal in k in the presence of interactions. In what follows we shall assume for simplicity that the parameter k does in fact refer to a linear momentum, and that momentum is conserved in the interaction. In this case the functions $G^{\pm}$ will clearly vanish unless $k = k_0$. Anticipating the rather reasonable result (Eq. 3.42) that the propagator depends in general only on the time difference, we write

$$G^{\pm}(k, t; k_0, t_0) = G^{\pm}(k, t - t_0)\, \delta(k, k_0). \tag{3.31}$$

In the course of development we shall have to introduce several other propagation functions, defined slightly differently but related directly to these two. One of particular importance in our theory is the "causal propagator" $G^{c}(k, t - t_0)$, which describes the propagation of an excitation from an earlier to a later time; that is, it describes a particle propagation from t_0

to t if t is the later time, and a hole propagation from t to t_0 if t_0 is the later time:

$$G^c(k, t - t_0) = -iG^+(k, t - t_0) \quad (t - t_0 > 0)$$
$$= i\sigma G^-(k, t - t_0) \quad (t - t_0 < 0). \tag{3.32}$$

To put this in a single equation,

$$G^c(k, t - t_0) = -i\theta(t - t_0)\, G^+(k, t - t_0)$$
$$+ i\sigma\theta\,(t_0 - t)\, G^-(k, t - t_0), \tag{3.33}$$

where $\theta(x)$ is defined in Eq. (3.30). From Eqs. (3.12, 13) we can write G^c directly thus:

$$G^c(k, t - t_0) = -i\,\langle T\,[\mathbf{a}_k(t)\, \mathbf{a}_k^\dagger(t_0)]\rangle, \tag{3.34}$$

where the symbol T denotes a rearrangement of factors in the order of descending time arguments from left to right:

$$T\,[A(t_1)\, B(t_2)] = \theta\,(t_1 - t_2)\, A(t_1)\, B(t_2)$$
$$- \sigma\theta\,(t_2 - t_1)\, B(t_2)\, A(t_1). \tag{3.35}$$

The factor $-\sigma$ in the second term of Eq. (3.33) and of Eq. (3.35) introduces a minus sign for FD operators, which has the property of making $T[(A(t_1)\, B(t_1)]$ unambiguous if the two operators anticommute for $t_2 = t_1$. Notice that $G^c(k, t - t_0)$ is not defined for $t - t_0 = 0$, and in general has a discontinuity at that point. The definition (3.35) is such that the discontinuity reflects the lack of commutativity (anticommutativity) of $a_k(t)$ and $a_k^\dagger(t)$.

For noninteracting particles, the relations (3.23, 25, 33) may be used to give this expression for G^c:

$$G_0^c\,(k, t - t_0) = -i\, e^{-i\varepsilon_k(t - t_0)}$$
$$\times\ [\theta\,(t - t_0) f^+(\bar{\varepsilon}_k) - \sigma\theta\,(t_0 - t) f^-(\bar{\varepsilon}_k)]. \tag{3.36}$$

For zero temperature, this becomes (Eq. 3.29)

$$G_0^c(k, t - t_0) = -i\, e^{-i\varepsilon_k(t - t_0)}$$
$$\times\ [\theta\,(t - t_0)\, \theta(\bar{\varepsilon}_k) - \theta\,(t_0 - t)\, \theta\,(-\bar{\varepsilon}_k)]. \tag{3.37}$$

We also define, primarily for use in our finite temperature theory, an "anticausal" propagator $G^{\sim c}(k, t - t_0)$:

$$G^{\sim c}(k, t - t_0) = i\,\langle \tilde{T}\,(\mathbf{a}_k(t)\, \mathbf{a}_k^\dagger(t_0))\rangle \tag{3.38}$$
$$= i\theta\,(t_0 - t)\, G^+(k, t - t_0) - i\sigma\theta\,(t - t_0)\, G^-(k, t - t_0) \tag{3.39}$$
$$= [G^c(k, t_0 - t)]^*, \tag{3.40}$$

where the definition of $\tilde{T}$, a modification of (3.35), is clear from (3.39).

3.2 Spectral representation

The time dependence of our propagators can be studied explicitly in terms of the spectrum of the true Hamiltonian H; while this spectrum is not known explicitly in physical cases, we can nevertheless determine some of the properites of our propagators in this way, and often gain some insight into their physical significance. Using the form (2.81) for the Heisenberg operators $\mathbf{a}_k(t)$ and $\mathbf{a}_k^\dagger(t)$, and using also the closure identity

$$\Psi = \sum_\nu \Psi_\nu(\Psi_\nu^\dagger \Psi) \quad \text{(any vector } \Psi\text{)}, \tag{3.41}$$

we can express $G^+(k, t - t_0)$ thus:

$$\begin{aligned} G^+(k, t - t_0) &= \langle \mathbf{a}_k(t)\, \mathbf{a}_k^\dagger(t_0)\rangle \\ &= \mathcal{Q}^{-1} \sum_{\lambda,\nu} e^{-\beta \bar{E}_\nu}\, \Psi_\nu^\dagger e^{iHt} a_k e^{-iHt} \Psi_\lambda \\ &\quad \times \Psi_\lambda^\dagger\, e^{iHt_0} a_k^\dagger e^{-iHt_0} \Psi_\nu \\ &= \mathcal{Q}^{-1} \sum_{\lambda,\nu} e^{-\beta \bar{E}_\nu} e^{iE_\nu t} e^{-iE_\lambda(t-t_0)} e^{-iE_\nu t_0} |\Psi_\nu^\dagger\, a_k \Psi_\lambda|^2 \\ &= \mathcal{Q}^{-1} \sum_{\lambda,\nu} e^{-\beta \bar{E}_\nu} e^{-i\omega_{\lambda\nu}(t-t_0)} |\Psi_\nu^\dagger a_k \Psi_\lambda|^2, \end{aligned} \tag{3.42}$$

where

$$\omega_{\lambda\nu} = E_\lambda - E_\nu, \tag{3.43}$$

and the fact (Eq. 3.31) that G^+ depends only on the difference $t - t_0$ is now demonstrated.

The Fourier transform of $G^+(k, t)$ (the variable t_0 may now be taken as zero for most purposes) can be read off from Eq. (3.42) directly. We write

$$G^+(k, t) = \int_{-\infty}^{\infty} \varrho^+(k, \omega)\, e^{-i\omega t}\, d\omega, \tag{3.44}$$

with

$$\varrho^+(k, \omega) = \mathcal{Q}^{-1} \sum_{\lambda,\nu} \delta(\omega - \omega_{\lambda\nu})\, e^{-\beta \bar{E}_\nu}\, |\Psi_\nu^\dagger a_k \Psi_\lambda|^2. \tag{3.45}$$

The spectral function $\varrho^+(k, \omega)$ consists of a sum of delta functions for a discrete system, but is expected to go over into a continuous function of ω in the limit of infinite volume.

A completely similar derivation yields the corresponding relations for $G^-(k, t)$:

$$G^-(k, t) = \int \varrho^-(k, \omega)\, e^{-i\omega t}\, d\omega; \tag{3.46}$$

$$\varrho^-(k, \omega) = \mathcal{Q}^{-1} \sum_{\lambda,\nu} \delta(\omega - \omega_{\lambda\nu})\, e^{-\beta \bar{E}_\lambda}\, |\Psi_\nu^\dagger a_k \Psi_\lambda|^2. \tag{3.47}$$

(In deriving (3.47), the statistical average is arbitrarily taken over states Ψ_λ, rather than Ψ_ν, and the completeness sum (3.41) over the states Ψ_ν, in order that the expression for $\varrho^-(k, \omega)$ may differ in form from that for $\varrho^+(k, \omega)$ only in the factor $e^{-\beta \bar{E}_\lambda}$.) The above discussion is quite applicable to the case of the canonical ensemble, the differences being that the actual energies E_ν would normally be used instead of $\bar{E}_\nu$ (though there is no harm in using $\bar{E}_\nu$ since N is held fixed), and that the states summed over are restricted by $N_\nu = N$ in Eq. (3.45), and $N_\lambda = N$ in Eq. (3.47). However, part of the following discussion, and therefore much of what we do subsequently, is not immediately applicable to the case where N is held fixed.

It is useful to relate the two functions ϱ^+ and ϱ^- to a single spectral function ϱ by making use of the fact that $G^+(k, t)$ and $G^-(k, t)$ can be combined in a natural way to give the expectation value of a commutator or anti-commutator (Eqs. 3.12, 13, 31). Using the symbol σ defined in Eq. (3.22) to distinguish between FD and BE statistics, we have

$$\begin{aligned} \langle [\mathbf{a}_k(t), \mathbf{a}_k^\dagger(0)]_\sigma \rangle &= G^+(k, t) + \sigma G^-(k, t) \\ &= \int [\varrho^+(k, \omega) + \sigma \varrho^-(k, \omega)]\, e^{-i\omega t}\, d\omega \\ &= \int \varrho(k, \omega)\, e^{-i\omega t}\, d\omega, \end{aligned} \tag{3.48}$$

where

$$\varrho(k, \omega) = \varrho^+(k, \omega) + \sigma \varrho^-(k, \omega) \tag{3.49}$$

$$= \mathcal{Q}^{-1} \sum_{\lambda,\nu} \delta(\omega - \omega_{\lambda\nu})\, (e^{-\beta \bar{E}_\nu} + \sigma e^{-\beta \bar{E}_\lambda})\, |\Psi_\nu^\dagger a_k \Psi_\lambda|^2. \tag{3.50}$$

Note that both ϱ^+ and ϱ^- are everywhere nonnegative, as is ϱ itself in the FD case, or $\bar{\omega}\varrho$ in the BE case. (We use $\bar{\omega}$ to represent $\omega - \mu$, in a similar manner to Eq. (3.21).) The commutation relations (2.21, 55) can be used with Eq. (3.48) in the special case $t = 0$ to give a normalization relation for the spectral function:

$$\int_{-\infty}^{\infty} \varrho(k, \omega)\, d\omega = 1. \tag{3.51}$$

We can also express $\varrho^{\pm}(k, \omega)$ in terms of $\varrho(k, \omega)$, if we note first that

$$\frac{e^{-\beta \bar{E}_\nu}}{e^{-\beta \bar{E}_\nu} + \sigma e^{-\beta \bar{E}_\lambda}} = \frac{1}{1 + \sigma e^{-\beta \bar{\omega}_{\lambda\nu}}}$$

$$= f^{+}(\bar{\omega}_{\lambda\nu}), \tag{3.52}$$

and that

$$\frac{e^{-\beta \bar{E}_\lambda}}{e^{-\beta \bar{E}_\nu} + \sigma e^{-\beta \bar{E}_\lambda}} = \frac{1}{e^{\beta \bar{\omega}_{\lambda\nu}} + \sigma}$$

$$= f^{-}(\bar{\omega}_{\lambda\nu}). \tag{3.53}$$

(see Eqs. 3.27, 20). (Here

$$\bar{\omega}_{\lambda\nu} = \bar{E}_\lambda - \bar{E}_\nu = \omega_{\lambda\nu} - \mu, \tag{3.54}$$

because, in the sums over states we are interested in the state Ψ_ν always has one fewer particles than the state Ψ_λ.) So, because of the factor $\delta(\omega - \omega_{\lambda\nu})$ in the sums, we have (for the ground state or grand ensemble only)

$$\varrho^{\pm}(k, \omega) = \varrho(k, \omega) f^{\pm}(\bar{\omega}), \tag{3.55}$$

and so

$$G^{\pm}(k, t) = \int \varrho(k, \omega) f^{\pm}(\bar{\omega}) e^{-i\omega t} d\omega. \tag{3.56}$$

The spectral representation of the causal propagator is easily obtained from Eqs. (3.33, 44, 46, 56):

$$G^{c}(k, t) = -i \int [\theta(t) \varrho^{+}(k, \omega) - \sigma \theta(-t) \varrho^{-}(k, \omega) e^{-i\omega t} d\omega \tag{3.57}$$

$$= -i \int \varrho(k, \omega) [\theta(t) f^{+}(\bar{\omega}) - \sigma \theta(-t) f^{-}(\bar{\omega})] e^{-i\omega t} d\omega. \tag{3.58}$$

As we have seen (Eq. 3.29), the weighting functions $f^{\pm}(\bar{\omega})$ become nonoverlapping step functions in the zero-temperature limit, so that in this limit the functions $\varrho^{\pm}(k, \omega)$ become independent, with nonoverlapping domains $\omega > \mu$ and $\omega < \mu$ respectively.

3.3 *Fourier transforms of propagators*

We introduce quite generally the Fourier transform of a propagator of any sort:

$$G^{(\ )}(k, t) = (2\pi)^{-1} \int G^{(\ )}(k, \omega) e^{-i\omega t} d\omega; \tag{3.59}$$

$$G^{(\ )}(k, \omega) = \int G^{(\ )}(k, t) e^{i\omega t} dt. \tag{3.60}$$

(The Fourier inversion theorem can be used freely if we think of our propagators as being generalized functions [48], or distributions.) For $G^{\pm}$, we clearly have (Eq. 3.56)

$$G^{\pm}(k, \omega) = 2\pi \varrho (k, \omega) f^{\pm}(\bar{\omega}), \tag{3.61}$$

and for G^c we get, using Eqs. (3.58, 60)

$$G^c(k, \omega) = -i \int e^{i\omega t} dt \int e^{-i\omega' t} d\omega' \varrho (k, \omega')$$
$$\times \; [\theta(t) f^{+}(\bar{\omega}') - \sigma \theta(-t) f^{-}(\bar{\omega}')] . \tag{3.62}$$

The t integration yields the Fourier transform of $\theta(t)$, given formally by

$$\int \theta(t)\, e^{iyt}\, dt = \int_0^{\infty} e^{iyt}\, dt . \tag{3.63}$$

This simple but important function is an analytic function of y, namely i/y, for y in the upper half plane ($\mathrm{Im}\, y > 0$) where the integral converges absolutely, but it is singular for real y—specifically at $y = 0$. The Fourier transform for real y is a generalized function (or Schwartz distribution) [48], and may be thought of as the boundary value [49], as one approaches the real axis from above, of the analytic function i/y:

$$\int \theta(t)\, e^{iyt}\, dt = \lim_{\eta \to 0+} (y + i\eta)^{-1}. \tag{3.64}$$

We shall consistently use η to represent a positive infinitesimal without displaying the limit explicitly. Indeed the limit exists only as a generalized function, or alternatively, may be taken after integration when the function appears in an integrand. The limiting behavior of the function $(y + i\eta)^{-1}$ is

$$(y + i\eta)^{-1} = y^{-1} - i\pi \delta (y), \tag{3.65}$$

where y^{-1}, for real y, shall indicate that the principal value is to be taken in any integration. This combination of principal value plus delta function appears frequently. The validity of Eq. (3.65) can be seen by looking at the real and imaginary parts of the left side in the limit of small η. The real part is $y/(y^2 + \eta^2)$, which behaves like $1/y$ except for $y \lesssim \eta$, where the function is cut off in a way that duplicates the principal value procedure. The imaginary part is $-\eta/(y^2 + \eta^2)$, which is a Lorentzian of infinitesimal width, but with integral equal to $-\pi$; it thus has the character of a delta function multiplied by $-\pi$. In summary, we thus can write

$$\int \theta(t)\, e^{iyt}\, dt = i (y + i\eta)^{-1} \tag{3.66}$$

$$= iy^{-1} + \pi \delta (y), \tag{3.67}$$

so that

$$G^c(k, \omega) = \int \varrho\,(k, \omega')\left[\frac{f^+(\bar\omega')}{\omega - \omega' + i\eta} + \sigma\,\frac{f^-(\bar\omega')}{\omega - \omega' - i\eta}\right] d\omega' \qquad (3.68)$$

$$= \int \varrho\,(k, \omega')\left[\frac{f^+(\bar\omega') + \sigma f^-(\bar\omega')}{\omega - \omega'}\right.$$

$$\left. - i\pi\,(f^+(\bar\omega') - \sigma f^-(\bar\omega'))\,\delta\,(\omega - \omega')\right] d\omega' \qquad (3.69)$$

$$= \mathscr{P}\int \frac{\varrho\,(k, \omega')}{\omega - \omega'}\, d\omega' - i\pi\alpha(\bar\omega)\,\varrho\,(k, \omega). \qquad (3.70)$$

Here use has been made of Eq. (3.26), and we have defined $\alpha(x)$ by

$$\alpha(x) = f^+(x) - \sigma f^-(x)$$
$$= \frac{e^{(1/2)\beta x} - \sigma\, e^{-(1/2)\beta x}}{e^{(1/2)\beta x} + \sigma\, e^{-(1/2)\beta x}} \qquad (3.71)$$

$$= (\tanh \tfrac{1}{2}\beta x)^\sigma. \qquad (3.72)$$

At zero temperature $\alpha(x)$ becomes the signum function:

$$\sigma(x) = \theta(x) - \theta(-x) = x/|x|. \qquad (3.73)$$

Yet another form for $G^c(k, \omega)$ can be obtained by noting that in Eq. (3.69) above, ω' can be replaced by ω in the argument of $f^\pm$, since in one term we have $f^+ + \sigma f^-$ which is simply unity, while in the second term the functions are multiplied by $\delta\,(\omega - \omega')$. We can therefore take the factors $f^\pm(\bar\omega)$ outside of the integral, and obtain

$$G^c(k, \omega) = f^+(\bar\omega)\, G^R(k, \omega) + \sigma f^-(\bar\omega)\, G^A(k, \omega), \qquad (3.74)$$

where

$$G^{R,A}(k, \omega) = \int \frac{\varrho\,(k, \omega')}{\omega - \omega' \pm i\eta}\, d\omega' \qquad (3.75)$$

are the "retarded" and "advanced" propagators, so called because their Fourier transforms can be shown to vanish for negative t and positive t respectively.

The Fourier transform of the anticausal propagator $G^{\sim c}(k, t)$ is easily obtained by using Eq. (3.40), from which it follows that

$$G^{\sim c}(k, \omega) = [G^c(k, \omega)]^* \qquad (3.76)$$

$$= f^+(\bar\omega)\, G^A(k, \omega) + \sigma f^-(\bar\omega)\, G^R(k, \omega). \qquad (3.77)$$

It will be useful later to have $G^{\pm}$, too, expressed in terms of the retarded and advanced functions. From Eq. (3.65) and its complex conjugate, and Eq. (3.75), we can see that

$$G^A(k, \omega) - G^R(k, \omega) = 2\pi i \varrho\,(k, \omega), \tag{3.78}$$

so that from Eq. (3.61) for $G^{\pm}(k, \omega)$, we can write

$$G^{\pm}(k, \omega) = if^{\pm}(\bar{\omega})\,[G^R(k, \omega) - G^A(k, \omega)]. \tag{3.79}$$

The retarded and advanced functions are the boundary values, approaching from above and below the real axis, of a function $G(k, z)$, analytic off the real axis, which we shall speak of as the "analytic propagator":

$$G(k, z) = \int \frac{\varrho\,(k, \omega')}{z - \omega'}\, d\omega' \quad (\mathrm{Qm}\, z \neq 0); \tag{3.80}$$

$$G^{R,A}(k, \omega) = G\,(k, \omega \pm i\eta). \tag{3.81}$$

The free-particle causal propagator $G_0^c(k, \omega)$ can be obtained directly by taking the Fourier transform of Eq. (3.36), in the same manner as above; we get

$$G_0^c(k, \omega) = \frac{f^+(\bar{\varepsilon}_k)}{\omega - \varepsilon_k + i\eta} + \sigma\,\frac{f^-(\bar{\varepsilon}_k)}{\omega - \varepsilon_k - i\eta}. \tag{3.82}$$

This can be written in alternative forms, analogous to Eqs. (3.70) and (3.74), and obtained by similar reasoning:

$$G_0^c\,(k, \omega) = (\omega - \varepsilon_k)^{-1} - i\pi\alpha(\bar{\omega})\,\delta\,(\omega - \varepsilon_k) \tag{3.83}$$

$$= \frac{f^+(\bar{\omega})}{\omega - \varepsilon_k + i\eta} + \sigma\,\frac{f^-(\bar{\omega})}{\omega - \varepsilon_k - i\eta}. \tag{3.84}$$

The zero-temperature form plays a particularly important rôle in the ground state expansion of the next chapter:

$$G_0^c(k, \omega) = \frac{\theta(\bar{\varepsilon}_k)}{\omega - \varepsilon_k + i\eta} + \frac{\theta(-\bar{\varepsilon}_k)}{\omega - \varepsilon_k - i\eta}$$

$$= [\omega - \varepsilon_k + i\eta\sigma\,(\bar{\varepsilon}_k)]^{-1} \quad (\beta \to \infty). \tag{3.85}$$

Here use has been made of the signum function (Eq. 3.73), and of the limiting forms (Eq. 3.29) of $f^{\pm}(\bar{\varepsilon}_k)$. If we use the same reasoning on Eq. (3.84) we find an alternative zero-temperature form:

$$G_0^c(k, \omega) = [\omega - \varepsilon_k + i\eta\sigma(\bar{\omega})]^{-1} \quad (\beta \to \infty). \tag{3.86}$$

The equivalence of (3.85) and (3.86) lies in the fact that the $i\eta$ only plays a rôle where the denominator vanishes, i.e., where $\omega = \varepsilon_k$. The form (3.86) stresses the property [Eqs. (3.74, 81, 29)] that at zero temperature the causal propagator consists of boundary values of the analytic propagator—from above the real axis for $\omega > \mu$ and from below for $\omega < \mu$.

Comparing the general form Eq. (3.56) for $G^{\pm}(k, t)$ with the specific forms [Eqs. (3.23, 25)] they takefor noninteracting particles, we see that the spectral function $\varrho\,(k, \omega)$ in that case becomes simply

$$\varrho_0(k, \omega) = \delta\,(\omega - \varepsilon_k), \tag{3.87}$$

from which any of the free-particle propagators can be obtained, using, for example, (3.75), (3.80), etc. Equation (3.87) could also have been obtained from the general expression (Eq. 3.50) for $\varrho\,(k, \omega)$; the details are a bit complicated, but it is clear that in the absence of interactions the matrix element $\Psi_\nu^\dagger a_k \Psi_\lambda$ vanishes unless the state Ψ_λ differs from the state Ψ_ν only by the presence of one additional particle in the single-particle state k. Thus $\omega_{\lambda\nu}$ has to be equal to ε_k, and the normalization is determined by (3.51).

This gives us some further insight into the meaning of $\varrho\,(k, \omega)$ in general. A delta-function component in ϱ gives rise to an undamped component in one of the propagators $G^{\pm}$, corresponding to a particle or to a particle-like behavior of the excitation. A shift of the position of such a delta function corresponds to a shift in energy of the excitation from its free-particle value. A peak of small but finite width in ϱ as a function of ω (we are speaking of an infinite medium now) would correspond [through Eq. (3.56)] to a damped propagation of the excitation; that is, it would have a finite lifetime for decay into other modes, such as collective modes, particle-hole pairs, or complicated states which cannot be simply described at all. If the width is small, one can speak of a "quasiparticle"—an excitation which is like a particle in having a definite or nearly definite energy for a given definite momentum. A broad spectral distribution for a given momentum, or a broad component, would arise from states of the system of a non-quasiparticle type. The absence of a narrow peak does not mean there cannot be quasiparticle-type excitations—it just means that such excitations do not enter into the matrix elements $\Psi_\nu^\dagger a_k \Psi_\lambda$ which determine the behavior of $\varrho\,(k, \omega)$; that is, only a quasiparticle excitation with the quantum numbers (or basic character) of a particle or a hole will give rise to a peak in $\varrho(k, \omega)$, while a quasiparticle of different character such as a collective phonon-type excitation, or a bound particle-hole pair (e.g. an exciton), would not appear here.

It is well here to consider the zero-temperature limit explicitly, since, while Eq. (3.50) describes this limit accurately, it does obscure some important features. As has been mentioned, the domains in ω over which $\varrho^+(k, \omega)$ and $\varrho^-(k, \omega)$ are nonzero become non-overlapping in this limit, and it is clearer to consider them separately, since ϱ becomes equal to ϱ^+ for $\omega > \mu$, and to $\sigma\varrho^-$ (the factor σ refers as usual to the choice of statistics) for $\omega < \mu$. For $\varrho^+(k, \omega)$, Eq. (3.45) indicates that the dominant term in the sum over states ν is the N-particle ground state $\Psi_{N,0}$, N having that value for which $\bar{E}_\nu$ is a minimum. In this limit, then, ϱ^+ is given by

$$\begin{aligned}\varrho^+(k, \omega) &= \sum_\lambda \delta\,[\omega - (E_\lambda - E_{N,0})]\,|\Psi^\dagger_{N,0} a_k \Psi_\lambda|^2 \\ &= \sum_\lambda \delta\,[\omega - (E_\lambda - E_{N,0})]\,|\Psi^\dagger_\lambda a^\dagger_k \Psi_{N,0}|^2,\end{aligned} \tag{3.88}$$

where the second expression emphasizes that what enters is the scalar product of the $(N+1)$-particle state Ψ_λ with the state in which one particle is added to the N-particle ground state. The state Ψ_λ clearly has an energy greater than $E_{N,0} + \mu$ (since $\bar{E}_\lambda > E_{N,0}$) so we do have $\omega > \mu$. For $\varrho^-(k, \omega)$, on the other hand, it is the term in Eq. (3.47) for which $\Psi_\lambda = \Psi_{N,0}$ which dominates, so that

$$\varrho^-(k, \omega) = \sum_\nu \delta\,[\omega + (E_\nu - E_{N,0})]\,|\Psi^\dagger_\nu a_k \Psi_{N,0}|^2, \tag{3.89}$$

and we are dealing now with the overlap between the $(N-1)$-particle states Ψ_ν and the state in which one particle is removed from the N-particle ground state. This indicates that a peak in $\varrho(k, \omega)$ for $\omega < \mu$ corresponds roughly to a hole type of excitation, particularly in the fermion case.

3.4 Macroscopic properties of the system

The equilibrium properties of the system viewed in the large can be obtained in principle from any one of several thermodynamic functions, two of which can in fact be obtained directly from the one-particle spectral function $\varrho(k, \omega)$, once that is known. We shall consider the temperature T (or its inverse β), the chemical potential μ and the total volume V as the independent thermodynamic variables, though the transformation to a different set (including the mean particle number $\mathcal{N}$ for example) is straightforward. The value of the mean particle number is given by the expectation value of the corresponding operator (Eq. 2.64):

$$\mathcal{N} = \langle \mathbf{N} \rangle = \sum_k \langle a_k^\dagger a_k \rangle$$

$$= \sum_k G^-(k, 0) \tag{3.90}$$

$$= \sum_k \int d\omega \, \varrho^-(k, \omega) \tag{3.91}$$

$$= \sum_k \int d\omega \, \varrho(k, \omega) f^-(\bar{\omega}), \tag{3.92}$$

where we have used Eqs. (3.13, 31, 46, 55) [note that the form (3.92) could not be used in the case of the canonical ensemble]. In the limit of very large volumes, the sum over k values can be replaced by an integral provided the summand is a well-behaved function of k in the limit, because the cell volume Δ_k in k-space, satisfying

$$\Delta_k = (2\pi)^3 V^{-1}, \tag{3.93}$$

becomes arbitrarily small, and from the definition of the Riemann integral,

$$\lim_{V\to\infty} (2\pi)^3 V^{-1} \sum_k (\;) = \lim_{\Delta_k \to 0} \sum_k (\;) \Delta_k = \int (\;) \, d^3k. \tag{3.94}$$

We see, then, the particle density is given by

$$\mathcal{N}/V = (2\pi)^{-3} \int d^3k \int d\omega \, \varrho^-(k, \omega). \tag{3.95}$$

The existence of the limit corresponds to the physical requirement that the infinite medium can be characterized by intensive properties. For a large finite sample, the approximation of passing from the sum to an integral corresponds to the neglect of surface effects and other effects due to finite size.

In the case of the two body interaction (Eq. 2.68), we can use the expression (2.84) for H, so that the internal energy is given by

$$E = \langle H \rangle = \tfrac{1}{2} \sum_k \langle \mathbf{a}_k^\dagger(0) \, [i\dot{\mathbf{a}}_k(0) + \varepsilon_k \mathbf{a}_k(0)] \rangle$$

$$= \tfrac{1}{2} \sum_k [(i \, d/dt + \varepsilon_k) \, G^-(k, t)]_{t=0} \tag{3.96}$$

$$= \tfrac{1}{2} \sum_k \int d\omega \, (\omega + \varepsilon_k) \, \varrho^-(k, \omega). \tag{3.97}$$

As in the case of $\mathscr{N}$, the infinite volume limit yields an expression for the energy density,

$$E/V = \tfrac{1}{2}(2\pi)^{-3} \int d^3k \int d\omega\, (\omega + \varepsilon_k)\, \varrho^-(k, \omega)$$
$$= \tfrac{1}{2}(2\pi)^{-3} \int d^3k \int d\omega\, (\omega + \varepsilon_k)\, \varrho\, (k, \omega) f^-(\bar{\omega}). \qquad (3.98)$$

This expression is valid only in the case of two-body forces, and bears a close relation to the Hartree–Fock expression for the energy of a system. The term in the integrand proportional to ε_k yields a contribution to E which is seen to be $\frac{1}{2}\langle H_0 \rangle$, that is, half the expectation value of the kinetic energy. The term proportional to ω may be thought of as giving the effective single-particle energy for each value of k; this includes the kinetic energy and an effective potential energy due to the other particles in the medium. In order not to count this interaction energy twice in summing over all the particles, we have to have the factor $\varepsilon_k + \frac{1}{2}(\omega - \varepsilon_k)$, which is equivalent to the factor $\frac{1}{2}(\omega + \varepsilon_k)$ which we have. If we had pure r-body forces (instead of two-body forces), the factor $\varepsilon_k + (\omega - \varepsilon_k)/r$ would appear, while similar but more complicated relations hold in the case of the electron-phonon type of interaction (2.72).

Now the quantities E and $\mathscr{N}$, regarded as functions of μ, T, and V, satisfy a thermodynamic relation,

$$\partial E/\partial\mu = T\, \partial\mathscr{N}/\partial T + \mu\, \partial\mathscr{N}/\partial\mu, \qquad (3.99)$$

which follows from the fact that E and $\mathscr{N}$ can both be obtained by means of partial differentiation of the grand partition function $\mathscr{Q}$ (Eq. 3.6). This imposes a restriction on the spectral function $\varrho\,(k, \omega)$ through Eqs. (3.92, 97), if it is to be consistent with the assumption of two-body forces. In principle then, $\mathscr{N}$ alone is sufficient for calculating thermodynamic properties, since E can be obtained from Eq. (3.99) and a knowledge of its behavior near zero density; in practice, however, it is useful to be able to calculate E directly from (3.98) to avoid the necessity of integrating with respect to μ. The entropy may be obtained from any of several thermodynamic relations, such as

$$T\, \partial S/\partial T = \partial\, (E - \mu\mathscr{N})/\partial T, \qquad (3.100)$$

together with

$$\partial S/\partial\mu = \partial\mathscr{N}/\partial T, \qquad (3.101)$$

where again μ, T, and V are the independent variables. The pressure in turn is given, for the infinite homogeneous medium, by

$$PV = TS + \mu\mathscr{N} - E, \qquad (3.102)$$

$\mu\mathscr{N}$ being equal to the familiar Gibbs function.

If in a particular problem Eq. (3.98) cannot be used, it may be more straightforward, rather than using Eq. (3.99), to calculate directly the thermodynamic potential $\Omega\,(\mu, T, V)$, which is equal to $-PV, (= A - \mu\mathcal{N}; A$ is the Helmholtz free energy), from the fact that

$$\partial\Omega/\partial\mu = -\mathcal{N}, \tag{3.103}$$

together with a knowledge of its low density behavior. All of the other thermodynamic functions can be obtained directly from $\Omega\,(\mu, T, V)$ by standard procedures.

4

Perturbation expansion—Ground state

OUR AIM in this section and the next is to derive perturbation expansions for the propagators of our theory, both at absolute zero and at finite temperatures, that is, to express such a propagator as a power series in the interaction strength associated with the term H_1 in the Hamiltonian. We shall not examine carefully the question of the convergence of such an expansion, though we shall mention some of the directions in which one can attempt to overcome the limitations of the expansion method. The perturbation expansion method is one of the most useful features of the field theory method, since it permits a meaningful analysis of different contributions even up to high orders. It is true that in high order the terms of the expansion become very complicated; one feature of the field theory method is that through the use of Wick's theorem [50, 51] this complicatedness expresses itself in the appearance of a large number of terms, each of which is of a fairly simple general form. These terms, in fact, are shown to be in one-to-one correspondence with a set of diagrams, or Feynman graphs, drawn according to certain basic rules; the component parts of each diagram correspond in a neat way to the factors in the corresponding term in the expansion, so that the topological characteristics of the graph express pictorially the mathematical characteristics of the expansion terms. This often permits a more or less physical interpretation of the individual terms, and may allow one to argue that certain classes of terms are important for a given effect, and others unimportant. If, as sometimes happens, one can sum some infinite class of terms which one believes to be important, for example by solving a certain integral equation, results may be obtained which in some sense go beyond the perturbation expansion. It is virtually impossible, of course, to assess rigorously the validity of any such approximation procedure.

It should be well understood that the application of the Feynman rules to a particular problem is completely separate from the problem of deriving them, and is often much easier in some respects.

4.1 *Interaction representation—Basic expansion*

We shall work with operators in the interaction representation, defined by Eq. (2.85), and we shall find that the time-development operator $U(t, t_0)$, which proves basic in our analysis, is related in a simple way to the operator. which we call $U(t)$, which transforms from the Heisenberg to the interaction representation. Using Eqs. (2.81, 85), we find that the Heisenberg operator $\mathbf{A}(t)$ can be related to the interaction representation operator $A(t)$ by

$$\mathbf{A}(t) = e^{iHt} A\, e^{-iHt} = e^{iHt} e^{-iH_0 t} A(t)\, e^{iH_0 t} e^{-iHt}$$

$$= U^\dagger(t)\, A(t)\, U(t), \tag{4.1}$$

where the unitary operator $U(t)$ is given by

$$U(t) = e^{iH_0 t} e^{-iHt}. \tag{4.2}$$

(This cannot be written, even formally, as $e^{i(H_0 - H)t}$, since H and H_0 do not commute in general, and the power series expansions, for instance, would be different; $U(t)$ does have the property, however, that for $H_1 = 0$, $U(t) \equiv 1$.) It is clear that $U(t)$ is unitary if t is real:

$$U^\dagger(t)\, U(t) = (e^{iHt} e^{-iH_0 t})\,(e^{iH_0 t} e^{-iHt}) = 1; \tag{4.3}$$

and also that $U(t)$ satisfies the differential equation

$$i\, dU(t)/dt = e^{iH_0 t}(H - H_0)\, e^{-iHt} = H'(t)\, U(t), \tag{4.4}$$

with the boundary condition

$$U(0) = 1. \tag{4.5}$$

Here $H'(t)$ is the interaction Hamiltonian H_1 expressed in the interaction representation:

$$H'(t) = e^{iH_0 t} H_1\, e^{-iH_0 t}. \tag{4.6}$$

The development of the state vector of the system with time can also be expressed usefully in the interaction representation. We here let $\Psi_S(t)$ be the state vector in the conventional Schrödinger representation, satisfying (2.73):

$$i\, d\Psi_S(t)/dt = H\Psi_S(t), \tag{4.7}$$

and we define the state vector in the interaction representation by the same unitary transformation (Eq. 2.85) as that which defines operators in the interaction representation:

$$\Psi_I(t) = e^{iH_0t}\Psi_S(t). \tag{4.8}$$

This has roughly the effect of removing that part of the time dependence which is due to H_0, so that if the interaction is small or absent, $\Psi_I(t)$ is slowly varying or constant

$$i\, d\Psi_I(t)/dt = e^{iH_0t}(H - H_0)\Psi_S(t) = H'(t)\Psi_I(t) \tag{4.9}$$

Notice that we have divided the time dependence between the operators and the state vector in such a way that the behavior of the operators in the interaction representation (Eq. 2.86) is governed by H_0 while that of the state vector (Eq. 4.9) is governed by $H'(t)$. This means that the time dependence of the operators is known explicitly, while the development of the state vector is more amenable to expansion in powers of the interaction strength.

The time development of the state vector $\Psi_I(t)$ can be displayed formally using Eqs. (2.75) and (4.8):

$$\Psi_I(t) = e^{iH_0t}e^{-iHt}\Psi_I(0) = U(t)\Psi_I(0), \tag{4.10}$$

which also follows from (4.4, 5, 9). We have used the fact that at $t = 0$ both representations are the same [see Eq. (4.8)]. Furthermore, we can express $\Psi_I(t)$ in terms of the state at an arbitrary time t_0:

$$\Psi_I(t) = U(t)\, U^\dagger(t_0)\,\Psi_I(t_0) \tag{4.11}$$

$$= U(t, t_0)\,\Psi_I(t_0), \tag{4.12}$$

where

$$U(t, t_0) = U(t)\, U^\dagger(t_0). \tag{4.13}$$

Here we have used the unitarity property of $U(t)$ to express $\Psi_I(0)$ in terms of $\Psi_I(t_0)$, for some arbitrary time t_0, by means of Eq. (4.10). We have also defined a more general unitary operator $U(t, t_0)$, which describes the time development between two arbitrary times, and which, as noted previously, plays an important rôle in what follows; it is, in fact, the expansion of $U(t, t_0)$ in powers of the interaction strength which we use as the basis for our perturbation theory. The unitarity of $U(t, t_0)$ follows from the unitarity of $U(t)$; it is also clear that

$$U^\dagger(t, t_0) = U(t_0, t). \tag{4.14}$$

Since the second factor, $U^\dagger(t_0)$, in the right side of Eq.(4.13) is independent of t, it follows that $U(t, t_0)$ satisfies the same differential equation (Eq. 4.4) as does $U(t)$:

$$i\,\partial U(t, t_0)/\partial t = H'(t)\,U(t, t_0). \tag{4.15}$$

In this case the boundary condition is that

$$U(t_0, t_0) = 1. \tag{4.16}$$

We have seen (Eq. 4.1) how to relate the Heisenberg operators to the simpler interaction representation operators using $U(t)$. To reduce the problem of evaluating a propagator such as $G^c(k, t)$ (Eq. 3.34) to that of evaluating $U(t, t_0)$, it is necessary also to find an expression for the true ground state vector $\Psi_{N,0}$ with respect to which the expectation value is taken in Eq. (3.34), or in similar expressions for the other propagators. The basic idea is to start, at a time T in the distant past, with the unperturbed N-body ground state and the unperturbed Hamiltonian, and then to allow the Hamiltonian to change adiabatically—that is, very slowly over a very long time interval—to the true Hamiltonian. In the limit of infinitely slow change—the "adiabatic limit"—the final state of the system is the true ground state, by the adiabatic theorem, and so the development of the state from unperturbed to true ground state is described by the time-development operator $U(t, T)$. The expressions for $U(t, t_0)$ were derived on the assumption that the Hamiltonian was constant, but this restriction is easily relaxed; the explicit form for $U(t)$ given in Eq. (4.2) no longer holds if H_1 is time-dependent, but the differential equation (4.4), together with the boundary condition (4.5), still determines a unitary operator $U(t)$ which relates the Heisenberg to the interaction representation (Eqs. 4.1, 10). The Heisenberg representation itself is defined quite generally by Eq. (2.82), although (2.81) must be replaced by a more general unitary transformation. The operator $U(t, t_0)$ which describes the time development in the interaction representation (Eq. 4.12) still satisfies the relations (4.13–16).

What we would like to say, roughly, is that the true ground state vector is given by $U(0, -\infty)\,\Phi_{N,0}$, where $\Phi_{N,0}$ is the unperturbed ground state vector. This is not quite possible, since the shift in energy of the ground state gives rise to a factor of the form $e^{-i(\Delta E)t}$, where t, the time elapsed, becomes infinite in the limit. What has been shown [52] by more sophisticated arguments is that the true ground state $\Psi_{N,0}$ is given by

$$\Psi_{N,0} = c \lim \frac{U(0, -\infty)\,\Phi_{N,0}}{\Phi^\dagger_{N,0} U(0, -\infty)\,\Phi_{N,0}}, \tag{4.17}$$

where the limit is the adiabatic limit in which the interaction is turned on arbitrarily slowly; in this expression the denominator cancels the divergent phase factor, and the constant c is used to normalize $\Psi_{N,0}$ to unity. The time $t = 0$ is taken as a handy reference point, since the three representations (Heisenberg, interaction, and Schrödinger) coincide at that time.

Equation (4.17) can clearly be written in terms of $U(-\infty)$ by using Eq.(4.13):

$$\Psi_{N,0} = c \lim \frac{U^{\dagger}(-\infty)\,\Phi_{N,0}}{\Phi^{\dagger}_{N,0}U^{\dagger}(-\infty)\,\Phi_{N,0}}. \tag{4.18}$$

A similar expression in terms of $U(+\infty)$ can be obtained by requiring that the state vector become the unperturbed ground state vector at $t = +\infty$, after the interaction has been turned off adiabatically. Thus $\Psi_{N,0}$ is also given by

$$\Psi_{N,0} = c' \lim \frac{U^{\dagger}(+\infty)\,\Phi_{N,0}}{\Phi^{\dagger}_{N,0}U^{\dagger}(+\infty)\,\Phi_{N,0}}. \tag{4.19}$$

By using both of the equations (4.18, 19) to write the normalization condition for $\Psi_{N,0}$ we can obtain an expression for the product $c'^{*}c$, this being the quantity needed for our derivation:

$$\begin{aligned} c'^{*}c &= \lim \frac{\Phi^{\dagger}_{N,0}U(\infty)\,\Phi_{N,0}\Phi^{\dagger}_{N,0}U^{\dagger}(-\infty)\,\Phi_{N,0}}{\Phi^{\dagger}_{N,0}U(\infty)\;U^{\dagger}(-\infty)\,\Phi_{N,0}} \\ &= \lim \frac{\langle U(\infty)\rangle_0\,\langle U^{\dagger}(-\infty)\rangle_0}{\langle U\,(\infty,\,-\infty)\rangle_0}. \end{aligned} \tag{4.20}$$

The propagators with which we are concerned have the general typical form (Eqs. 3.12, 13)

$$G_{AB}(t, t_0) = \langle \mathbf{A}(t)\,\mathbf{B}(t_0)\rangle = \Psi^{\dagger}_{N,0}\,\mathbf{A}(t)\,\mathbf{B}(t_0)\,\Psi_{N,0}, \tag{4.21}$$

or else some combination of such expressions (Eqs. 3.33 ,34). Here $\mathbf{A}(t)$ and $\mathbf{B}(t_0)$ are any two Heisenberg operators, and the expectation value is taken with respect to the true ground state. It is convenient at this point to restrict ourselves to the case in which $t > t_0$. Using Eqs. (4.1, 18–20), we can write this as follows:

$$\begin{aligned} &G_{AB}(t, t_0) \\ &= \lim \frac{\Phi^{\dagger}_{N,0}\,U(\infty)\;U^{\dagger}(t)\;A(t)\;U(t)\;U^{\dagger}(t_0)\;B(t_0)\;U(t_0)\;U^{\dagger}(-\infty)\,\Phi_{N,0}}{\Phi^{\dagger}_{N,0}\,U(\infty,\,-\infty)\,\Phi_{N,0}} \\ &= \lim \frac{\langle U(\infty,\,t)\;A(t)\;U(t,\,t_0)\;B(t_0)\;U(t_0,\,-\infty)\rangle_0}{\langle U(\infty,\,-\infty)\rangle_0}. \end{aligned} \tag{4.22}$$

The interaction now enters only through the operator $U(t, t_0)$, for which we now develop a perturbation expansion.

The expansion we shall obtain is closely akin to the standard time-dependent perturbation theory known as the method of variation of parameters. The method of derivation looks different, however, and consists of converting the differential equation (4.15, 16) into an integral equation, whose iterated solution is the desired expansion.

By integrating Eq. (4.15) from t_0 to t [replacing t by t_1 in Eq. (4.15)] and using the boundary condition (4.16), we obtain this integral equation for $U(t, t_0)$:

$$U(t, t_0) = 1 - i \int_{t_0}^{t} dt_1 \, H'(t_1) \, U(t_1, t_0). \tag{4.23}$$

If $U(t, t_0)$ is assumed to have a power series expansion in the interaction strength (in H', roughly), it is seen that the nth-order term on the left is obtained by substituting the $(n-1)^{th}$-order term in the integral on the right. This yields the required expansion of $U(t, t_0)$, in the form

$$\begin{aligned} U(t, t_0) &= 1 - i \int_{t_0}^{t} dt_1 H'(t_1) + (-i)^2 \int_{t_0}^{t} dt_1 \int_{t_0}^{t_1} dt_2 \, H'(t_1) \, H'(t_2) + \cdots \\ &= \sum_{n=0}^{\infty} (-i)^n \int_{t_0}^{t} dt_1 \int_{t_0}^{t_1} dt_2 \cdots \int_{t_0}^{t_{n-1}} dt_n \, H'(t_1) \, H'(t_2) \cdots H'(t_n) \\ &= \sum_{n=0}^{\infty} (-i)^n \iint \cdots \int dt_1 \, dt_2 \cdots dt_n \, H'(t_1) \, H'(t_2) \cdots H'(t_n) \\ & \qquad (t_0 < t_n < t_{n-1} < \cdots < t_1 < t). \end{aligned} \tag{4.24}$$

(The $n = 0$ term is understood to be unity.) Very little is known [53] about the convergence of this expansion for the many-particle system, though we expect it to be useful, at any rate, if the interaction is weak or the density is low.

It is very useful to express this as an integral without restriction on the ordering of the variables t_i, so that each variable is integrated from t_0 to t. The effect of the ordering in Eq. (4.24) is to put the operators $H'(t_i)$ in order of descending argument from left to right; the same effect can be achieved artificially by using the ordering operator T, defined for two operators by Eq. (3.35), and here defined by

$$T[A_1(t_1) \, A_2(t_2) \cdots A_n(t_n)] = (-1)^P A_{i_1}(t_{i_1}) \, A_{i_2}(t_{i_2}) \cdots A_{i_n}(t_{i_n}), \tag{4.25}$$

where $(i_1, i_2, \ldots, i_n)$ is that permutation of the numbers $(1, 2, \ldots, n)$ which makes

$$t_{i_1} > t_{i_2} > \cdots > t_{i_n}, \tag{4.26}$$

and $(-1)^P$ indicates the parity (even, in the case we are considering at the moment) of the permutation of fermion operators induced by T. That is to say, the T-ordering operator simply rearranges the operators from left to right in order of descending argument t. If the operators A should happen to commute (anticommute), of course, then this becomes trivial. The limiting case in which some of the t's are equal is usually of no especial significance. If the equal-time operators should happen not to commute, one would expect to use a limiting procedure to avoid ambiguity.

If one introduces this ordering convention, then, one can dispense with the restriction on the order of the t's, and we have

$$U(t, t_0) = \sum_{n=0}^{\infty} (-i)^n (n!)^{-1} \int_{t_0}^{t} dt_1 \int_{t_0}^{t} dt_2 \cdots \int_{t_0}^{t} dt_n T[H'(t_1) H'(t_2) \cdots H'(t_n)]. \tag{4.27}$$

The factor $(n!)^{-1}$ is necessary because each of the $n!$ possible orders in which the variables t_i may appear now gives a contribution to the integral equal to the original integral appearing in Eq. (4.24). Note that the integrand is now symmetrical under interchange or permutation of the arguments t_i. The advantage of the T-ordered form lies in the fact that all the variables t_i are now integrated independently over the same range from t_0 to t.

We now wish to apply the expansion of $U(t, t_0)$ (Eq. 4.24 or 27) to the calculation of our prototype propagator (Eq. 4.22). The way in which the form of expansion of $U(t, t_0)$ depends on the fact that $t > t_0$ is what makes it particularly convenient in the propagator (4.22) also to restrict our attention to the case $t > t_0$. This, too, is why it is particularly appropriate, as we shall see later on, to calculate $G^c(k, t - t_0)$ (Eq. 3.34) rather than one of the other propagators such as $G^{\pm}(k, t - t_0)$.

We now insert the original expansion (Eq. 4.24) for $U(t, t_0)$ into the numerator in Eq. (4.22), to obtain

$$\begin{aligned} G_{AB}(t, t_0) = \langle U(\infty, -\infty)\rangle_0^{-1} & \sum_{k,l,m=0}^{\infty} (-i)^{k+l+m} \int (dt)^{k+l+m} \\ & \times \langle H'(t_1) \cdots H'(t_k) A(t) H'(t_{k+1}) \cdots H'(t_{k+l}) \\ & \times B(t_0) H'(t_{k+l+1}) \cdots H'(t_{k+l+m})\rangle_0, \end{aligned} \tag{4.28}$$

with the restriction on the integration variables t_i that

$$t_1 > t_2 > \cdots > t_k > t > t_{k+1} > \cdots > t_{k+l} > t_0 > t_{k+l+1} > \cdots > t_{k+l+m}. \tag{4.29}$$

We can easily combine all the nth-order terms—that is, those for which

$$k + l + m = n, \tag{4.30}$$

by eliminating the reference to t and t_0 in the condition (4.29), and using the T-ordering operator to place the factors $H'(t_i)$ in their proper position with respect to $A(t)$ and $B(t_0)$. Thus

$$G_{AB}(t, t_0) = \langle U(\infty, -\infty)\rangle_0^{-1} \sum_{n=0}^{\infty} (-i)^n \int_{-\infty}^{\infty} dt_1 \cdots \int_{-\infty}^{\infty} dt_n \times \langle T[H'(t_1)\, H'(t_2) \cdots H'(t_n)\, A(t)\, B(t_0)]\rangle_0, \tag{4.31}$$

with the simpler restriction now that

$$t_1 > t_2 > \cdots > t_n. \tag{4.32}$$

The different values of k, l, and m in Eq. (4.28) correspond respectively to the different numbers of arguments t_i which exceed t, lie between t_0 and t, or are less than t_0. We can now drop the restriction (4.32) on the arguments t_i also, by allowing the T operator in Eq. (4.31) to place the operators $H'(t_i)$ in their proper order with respect to each other. As in Eq. (4.27), we must compensate, with a factor $1/n!$, for the fact that the $n!$ different orders in which the t_i's may fall all give equal contributions to the integral. Thus we can write

$$G_{AB}(t, t_0) = \langle U(\infty, -\infty)\rangle_0^{-1} \sum_{n=0}^{\infty} (-i)^n (n!)^{-1} \int_{-\infty}^{\infty} dt_1 \cdots \int_{-\infty}^{\infty} dt_n \times \langle T[H'(t_1)\, H'(t_2) \cdots H'(t_n)\, A(t)\, B(t_0)]\rangle_0. \tag{4.33}$$

This expansion of the numerator is now very similar in form to the expression which one obtains for the denominator of our propagator by using Eq. (4.27),

$$\langle U(\infty, -\infty)\rangle_0 = \sum_{n=0}^{\infty} (-i)^n (n!)^{-1} \int_{-\infty}^{\infty} dt_1 \cdots \int_{-\infty}^{\infty} dt_n \langle T[H'(t_1) \cdots H'(t_n)]\rangle_0, \tag{4.34}$$

and differing from it only in the additional factors $A(t)$ and $B(t_0)$ in the integrand. Finally, we can apply the result (4.33) to the causal propagator $G^c(k, t - t_0)$ (Eq. 3.34), by simply inserting the operators $a_k(t)$ and $a_k(t_0)$ in place of $A(t)$ and $B(t_0)$, and allowing the T-ordering operator to arrange them in their proper order, with the proper sign in the case of fermion operators:

$$G^c(k, t - t_0) = -i < U(\infty, -\infty)\rangle_0^{-1} \sum_{n=0}^{\infty} (-i)^n (n!)^{-1} \int_{-\infty}^{\infty} dt_1 \cdots dt_n \langle T[H'(t_1) \cdots H'(t_n)\, a_k(t)\, a_k^\dagger(t_0)]\rangle_0. \tag{4.35}$$

Up to Eqs. (4.33, 34) our perturbation theory has been completely general, with no assumptions about the form of H_0 and H_1, and no obvious advantage over other expansion procedures. We have mentioned, however, the importance of Wick's theorem in sorting out even the high-order terms of the expansion into many fairly simple terms whose relative importance may be estimated on more or less physical grounds. The theorem is applicable in general to a T-ordered product of annihilation and creation operators in the interaction representation (where they behave, as we have seen, like free-particle operators), and takes a somewhat simpler form for the ground-state expectation value of such a product. The form of our perturbation expansion has thus been chosen to enable us to make use of this theorem.

4.2 *Wick's theorem and Feynman graphs*

There are certain complications in connection with the boson system, which arise because Wick's theorem applies in its basic form only when the expectation value is taken with respect to the vacuum state, or to a vacuum-like state such as the fermion ground state [Eqs. (2.59, 60) and discussion following]. In order to present the procedure without complications, therefore, we shall restrict our attention for the moment to fermions, and to bosons such as phonons (Eq. 2.72) which are absent in the unperturbed ground state, and thus give no trouble. Our primary emphasis here will be on the fermion problem, described by the interaction Hamiltonian (Eq. 2.68); a different form for H_1 would not alter the basic procedure, and indeed is necessary for bosons with no particles present in the unperturbed ground state, because virtual production would have to take place before the particles could interact. When we come to treat the many-boson problem, we shall find that we can consider an equivalent problem of just this type, that is, with a vacuum-like state for the unperturbed ground state and with H_1 describing production and annihilation processes as well as scattering processes.

We consider, then, the substitution of the interaction Hamiltonian (Eq. 2.68) into the propagator expansion (Eq. 4.35). Recalling that $H'(t)$ is just H_1 transformed to the interaction representation (Eq. 4.6), we see that what is involved in Eq. (4.35) is simply the time-ordered product of boson or fermion annihilation and creation operators in the interaction representation. The four operators in a single factor $H'(t_i)$ are all associated with the same time t_i and so the time-ordering operator does not tell how to order them. We want the creation operators $a_k^\dagger$ to the left, however, and so prescribe an argument $t_i + \varepsilon$ for all creation operators, with ε a positive infinitesimal. In most cases we shall find it unnecessary, but it does no harm, and will be there to help when needed. The fact that the two annihilation operators (or the two creation operators) have the same time argument is of no concern, since they commute (BE) or anticommute (FD), while the operator introduces a minus sign on interchange in the FD case, so that only the order in which they are written matters, and the presumed order of their time arguments is immaterial.

The form (not the most general form, in fact) in which we shall use Wick's theorem [50] is this: let $b_1, b_2, \ldots, b_{2p}$ be boson or fermion annihilation or creation operations in the interaction representation, associated with times $t_1, t_2, \ldots, t_{2p}$, respectively, and let $\langle(\ldots)\rangle_0$ represent an expectation value with respect to the unperturbed ground state Φ_0, which is either the boson vacuum or the N-fermion ground state. Then the theorem can be written schematically thus:

$$\langle T(b_1 b_2 \cdots b_{2p})\rangle_0 = \sum_{\text{a.p.p}} (-1)^P \prod \langle T(b_r b_s)\rangle_0 . \tag{4.36}$$

The sum is over all possible pairings (a.p.p.) of the factors b_r, which means all possible distinct ways of dividing them into p pairs. The product is over the p pairs, b_r and b_s represent any of the pairs in this product, and P has the parity of the permutation of fermion operators needed to bring the pairs together in the order in which they are written, starting from the order on the left side of the equation. This can be stated somewhat more precisely and explicitly by summing over permutations P of the form

$$(1, 2, \ldots, 2p) \to (r_1, r_1', r_2, r_2', \ldots, r_p, r_p'), \tag{4.37}$$

with the restrictions

$$\begin{aligned} r_1 < r_2 < \cdots < r_p; \\ r_s < r_s' \quad (s = 1, 2, \ldots, p). \end{aligned} \tag{4.38}$$

Then the theorem states that

$$\langle T(b_1 b_2 \cdots b_{2p})\rangle_0 = \sum_P (-1)^P \prod_{s=1}^{p} \langle T(b_{r_s} b_{r_s'})\rangle_0 . \tag{4.39}$$

The restrictions (4.38) are not in themselves essential, but are necessary in some form to ensure that each distinct pairing be counted just once. The theorem is proved in Appendix B.

One may understand this theorem as saying that we get the right answer by supposing that each of the excitations (particles or holes) brought into virtual existence by the creation operators propagates freely until it is destroyed by one of the annihilation operators; each excitation can be destroyed by any one of the annihilation operators, and so we sum over all the possible ways of pairing the creation operators with the annihilation operators. The combination $\langle T(b_r b_s)\rangle_0$, which represents the free propagation of the excitation, is nonzero only if b_r and b_s refer to the same single-particle state, and if the earlier time is associated with the creation of the excitation (particle or hole) and the latter time with its annihilation.

We can choose the order of factors in which we write the pair in such a way that the resulting factor is simply of the form (Eq. 3.34, 37)

$$\langle T(a_k(t_r)\, a_k^\dagger(t_s))\rangle_0 \equiv iG_0^c(k, t_r - t_s). \tag{4.40}$$

The theorem states, too, that in the FD case the exclusion principle can be ignored in individual terms in this picture, provided one sums over all the different ways in which the excitations can be created and then absorbed. (This includes the exclusion principle for holes, which says that there cannot be two holes in the same single-particle state!) Note, however, that the exclusion principle for the particles present in the ground state is always taken into account by the use of the hole picture, inasmuch as we never allow a negative number of holes to be present. To see how the terms which violate the exclusion principle can cancel each other, look, for example, at four fermion operators, all associated with the same single-particle state $k > N$, and with times t_1, t_2, t_3, and t_4, which we suppose to be already in order,

$$t_1 > t_2 > t_3 > t_4, \tag{4.41}$$

but all infinitesimal so we can ignore the factors $e^{-i\varepsilon_k t_i}$. There are a number of possible cases, of which we show some which are illustrative:

$$\begin{aligned}\langle a_1 a_2 a_3^\dagger a_4^\dagger\rangle_0 &= \langle T(a_1 a_2 a_3^\dagger a_4^\dagger)\rangle_0\\ &= \langle T(a_1 a_2)\rangle_0 \langle T(a_3^\dagger a_4^\dagger)\rangle_0 - \langle T(a_1 a_3^\dagger)\rangle_0 \langle T(a_2 a_4^\dagger)\rangle_0\\ &\quad + \langle T(a_1 a_4^\dagger)\rangle_0 \langle T(a_2 a_3^\dagger)\rangle_0\\ &= \langle a_1 a_2\rangle_0 \langle a_3^\dagger a_4^\dagger\rangle_0 - \langle a_1 a_3^\dagger\rangle_0 \langle a_2 a_4^\dagger\rangle_0 + \langle a_1 a_4^\dagger\rangle_0 \langle a_2 a_3^\dagger\rangle_0\\ &= 0 - 1 + 1 = 0;\end{aligned} \tag{4.42}$$

$$\begin{aligned}\langle a_1 a_2^\dagger a_3 a_4^\dagger\rangle_0 &= \langle a_1 a_2^\dagger\rangle_0 \langle a_3 a_4^\dagger\rangle_0 - \langle a_1 a_3\rangle_0 \langle a_2^\dagger a_4^\dagger\rangle_0 + \langle a_1 a_4^\dagger\rangle_0 \langle a_2^\dagger a_3\rangle_0\\ &= 1 - 0 + 0 = 1;\end{aligned} \tag{4.43}$$

$$\begin{aligned}\langle a_1^\dagger a_2 a_3 a_4^\dagger\rangle_0 &= \langle a_1^\dagger a_2\rangle_0 \langle a_3 a_4^\dagger\rangle_0 - \langle a_1^\dagger a_3\rangle_0 \langle a_2 a_4^\dagger\rangle_0 + \langle a_1^\dagger a_4^\dagger\rangle_0 \langle a_2 a_3\rangle_0\\ &= 0 - 0 + 0 = 0.\end{aligned} \tag{4.44}$$

In (4.43, 44) the steps involving the T operator are omitted, since they follow the same pattern as in (4.42), and are trivial since we have restricted all terms to descending time sequences. In each case, the nonzero terms or factors represent the different ways in which a particle can be created at an earlier time, propagate to a later time, and there be annihilated.

We must now try to keep clear the various factors that arise in applying this theorem to the expansion (4.35) of $G^c(k, t - t_0)$. Apart from trivial numerical factors there are just two types of factor: matrix elements of the form $v(r, s; r', s')$ [short for $v(k_r, k_s; k_{r'}, k_{s'})$] from the interaction Hamiltonian (Eq. 2.68), and propagators $G_0^c(k, t_i - t_j)$ from the various pairs arising from Wick's theorem. There are integrations to perform, over the different time variables, and summations over the different k variables. In each order of the expansion there is a number of terms coming from the different ways of pairing the a_k's and $a_k^\dagger$'s; these different pairings can be characterized by the ways in which they interconnect the various factors H'. Since each factor H' itself contains four factors a_k or $a_k^\dagger$, it can be involved in as many as four different pairs. The character of these interconnections can be described pictorially by connecting points with lines—the points represent the factors H' and the two fixed factors $a_k(t)$ and $a_k^\dagger(t_0)$, while the lines represent the pairings. The resulting pictures are known as Feynman graphs [51], and the problem of enumerating the terms of the perturbation expansion is equivalent to that of enumerating these graphs, drawn according to definite rules. Each component of a graph corresponds to a certain factor in the corresponding term of the expansion according to a definite prescription, which we must now work out.

Rather than using a single point to represent an interaction $H'(t_i)$ in our graph, it is convenient for some purposes (though by no means necessary) to use a vertex of the form shown in Figure 1, in order to differentiate

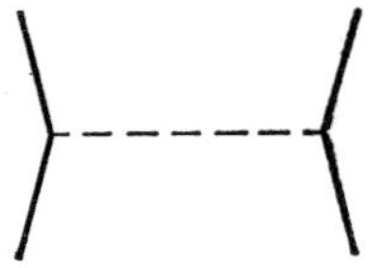

FIGURE 1 The vertex

between direct and exchange interactions. To start with, then, we draw for an nth-order graph n vertices of this form to represent the interactions $H'(t_i)$, and two additional points to represent the operators $a_k(t)$ and $a_k^\dagger(t_0)$. Each vertex and end-point is labeled with a time variable t_i $(i = 1, 2, \ldots, n)$, t, or t_0.

We now connect the vertices and endpoints with solid lines to represent the various pairs of factors a_k and $a_k^\dagger$ which are coupled together according to Wick's theorem; since each such factor $\langle T(a_r a_s^\dagger)\rangle_0$ is zero unless $k_r = k_s$, we can use a single momentum k_r to label each line. Each line can be given a direction, which we indicate by an arrow, since one end is associated with a particle creation operator and the other with a particle annihilation operator. We draw the arrow in the direction from the creation to the annihilation operator. If the factor $a_r^\dagger$ is at the earlier time, the arrow represents the propagation of a particle from the tail end to the head, while if the factor a_r is earlier the propagation of a hole is indicated, from the head to the tail of the arrow. (FD case only; this situation gives zero in the BE case.) In drawing the graphs and doing the related calculations, however, it is not necessary to distinguish between these two alternatives, and so we do not have to indicate the time ordering of the vertices in our graphs.

Our rules for drawing graphs are then as follows: for an nth-order graph there shall be n vertices of the form shown in Figure 1, labeled with the times t_i $(i = 1, 2, \ldots, n)$; there are two endpoints, labeled t and t_0. These vertices and endpoints are connected by directed lines, marked with an arrow and labeled by momenta k_r; two lines enter each vertex and two leave, one line

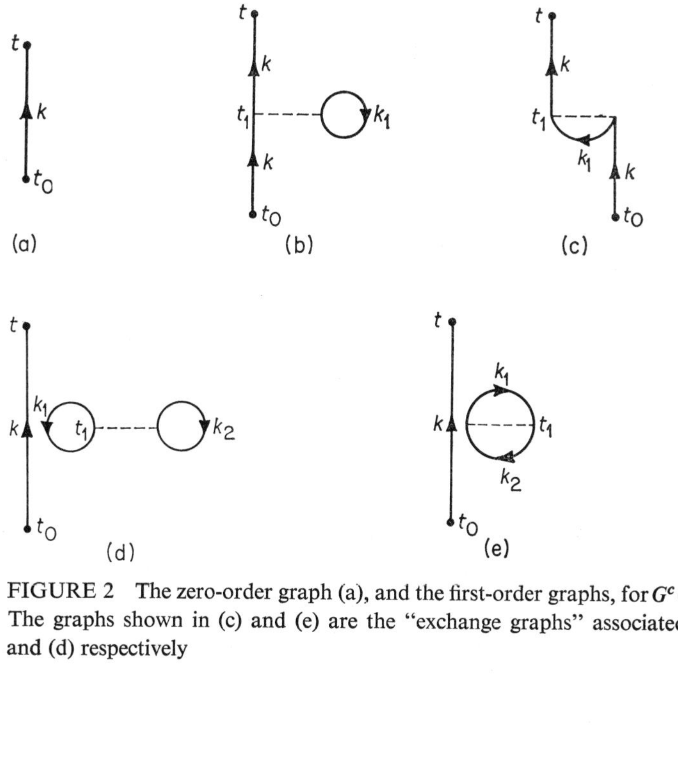

FIGURE 2 The zero-order graph (a), and the first-order graphs, for $G^c(k, t - t_0)$. The graphs shown in (c) and (e) are the "exchange graphs" associated with (b) and (d) respectively

has its head end at t, and one has its tail end at t_0. Those attached to the end-points are labeled with the momentum k, which is fixed, while the labels on all the others are dummy variables to be summed over, so that the choice of label is immaterial except for the ends. The zero- and first-order graphs are shown in Figure 2, and some of the second-order graphs are shown in Figure 3, with labels omitted.

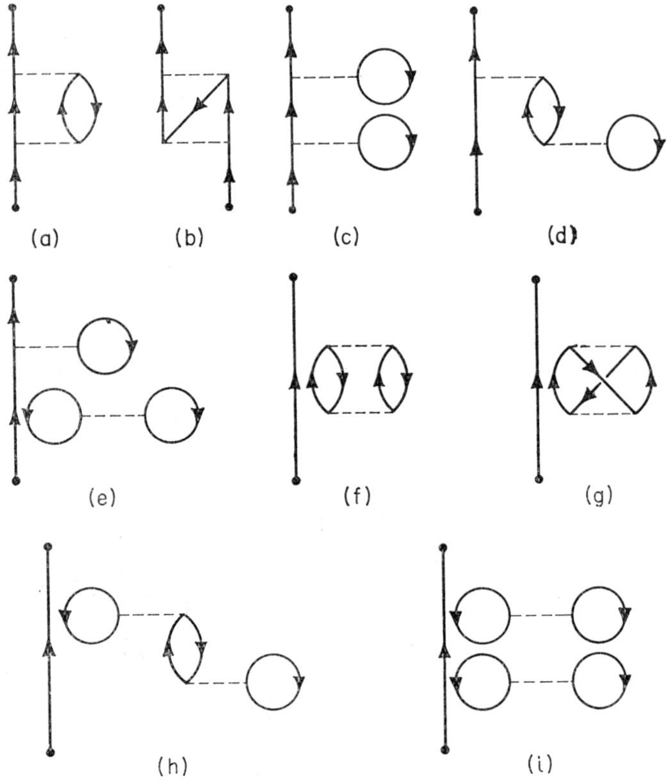

FIGURE 3 Second-order graphs for $G^c(k, t-t_0)$. Exchange graphs associated with (c), (d), (e), (h), and (i) are not shown (see Figure 2)

The factors associated with any graph can now be enumerated, by inspection of Eqs. (4.35, 39, 40; 2.68); this is not the final prescription, since further simplifications will be found possible. We have the overall factors $-i\langle U(\infty, -\infty)\rangle_0^{-1}$ and $(-1)^P(n!)^{-1}$, where the factor $(-1)^P$ will have to be studied further. For each vertex associated with time t_i, outgoing momenta k_r, k_s, and incoming momenta k_r', k_s', say, there is a factor

$$-\tfrac{1}{2} i\, v\,(r, s; r', s').$$

For each line, with momentum label k_r and running from time t_j to time t_i, there is a factor

$$iG_0^c\,(k_r, t_i - t_j).$$

The times t_i are to be integrated from $-\infty$ to $+\infty$, and the momenta k_r are to be summed over, apart from the momentum k which labels the lines attached to the endpoints. If a line begins and ends at the same vertex, the corresponding factor $G_0^c\,(k, t)$ must be evaluated in the limit that t approaches zero from negative values, since this corresponds to keeping the factor $a_k^\dagger$ to the left of the factor a_k (Eq. 3.34). This is the same as using $a_k(t_i)$ and $a_k^\dagger(t_i + \varepsilon)$ in $H'(t_i)$ as discussed earlier in this section.

Each graph represents in general several equivalent ways of connecting the factors a_k and $a_k^\dagger$ in pairs, for two reasons: one is the equivalence of the two ends of the dotted line at a vertex, which follows from the symmetry of the interaction matrix element (Eq. 2.69):

$$v\,(1, 2; 1', 2') = v\,(2, 1; 2', 1'); \tag{4.45}$$

the other is the complete equivalence of the different vertices, so that each way of assigning the times t_i to the various vertices in a graph yields the same contribution to $G^c(k, t - t_0)$, even though it represents a distinct way of picking the pairs of factors a_k and $a_k^\dagger$. Because there are $n!$ ways of assigning the variables t_i to the n vertices, we see that the factor $1/n!$ disappears, except in the case that a permutation of vertices may fail to yield a distinct graph, as can happen, for instance, in Figures 3(f) to 3(i). In such cases, which occur in this particular theory only in graphs with disconnected parts, the factor $1/n!$ is replaced by a factor not involving n, but depending only on the order of symmetry of the disconnected part.

Apart from the factor $\langle U(\infty, -\infty)\rangle_0^{-1}$, which we shall come back to, the graph in Figure 2(a) gives the zero-order contribution $G_0^c(k, t - t_0)$, while by the rules we have just worked out, Figure 2(b) yields the following:

$$-\,(-i)\,(-i)\,(i)^3 \sum_{k_1} v\,(k, k_1; k, k_1) \int_{-\infty}^{\infty} dt_1\, G_0^c\,(k, t - t_1)$$
$$\times\; G_0^c\,(k, t_1 - t_0)\; G_0^c\,(k_1, -\varepsilon).$$

The factor $\frac{1}{2}$ is dropped because the equivalent graph mentioned above and shown in Figure 4 has been included also. This will generally happen at every vertex, so that the factors $\frac{1}{2}$ may be dropped entirely, except in cases where because of the symmetry of the graph the interchange of the two ends

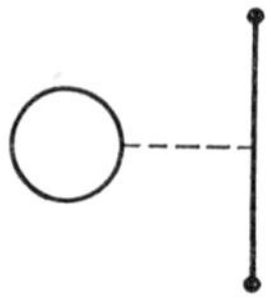

FIGURE 4 Graph equivalent to Figure 2(b)

of the vertex does not yield a different graph. This exception occurs in Figures 2(d, e), for example, and in many of the graphs with disconnected portions, but again does not occur in this theory in graphs without disconnected portions. The minus sign in front is the factor $(-1)^P$, which results from the fact that an odd permutation of fermion operators is needed to bring the factors from the form in which they are written in Eq. (4.35),

$$\langle v\,(k, k_1; k, k_1)\, T\,[a_k^\dagger(t_1)\, a_{k_1}^\dagger(t_1)\, a_{k_1}(t_1)\, a_k(t_1)\, a_k(t)\, a_k^\dagger(t_0)]\rangle_0,$$

to the form brought about by Wick's theorem (Eq. 4.36):

$$\langle T\,[a_k(t)\, a_k^\dagger(t_1)]\rangle_0\, \langle T\,[a_k(t_1)\, a_k^\dagger(t_0)]\rangle_0\, \langle T\,[a_{k_1}(t_1)\, a_{k_1}^\dagger(t_1 + \varepsilon)]\rangle_0.$$

(It is the order in which they are *written* here that counts—the further rearrangement induced by the T operators is all taken care of.) Quite generally, the factor $(-1)^P$ can be shown (this is straightforward, though we do not prove it here) to introduce a factor (-1) for each closed fermion loop that appears in a given graph, where a "closed loop" refers to a sequence of lines touching end to end to form a loop. The graphs in Figures 2(b, e) and 3(a, e, g, h) each contain an odd number of closed loops, while all the others in Figures 2 and 3 have an even number.

The denominator $\langle U(\infty, -\infty)\rangle_0$ of Eq. (4.35) is evaluated by exactly the same sort of procedure as we have worked out for the numerator. We simply apply the same arguments to Eq. (4.34), and we find that the only difference is in the absence of the endpoints from our graphs. The zero-order term is just unity, and the graphs associated with the first and second orders are shown in Figure 5. It will be noticed that these same graphs appear as subgraphs in some of the graphs of Figures 2 and 3, and indeed it is not hard to see that every graph lacking such disconnected portions (a "linked graph") such as Figures 2(a–c), 3(a–d), will also appear in higher orders with each of the graphs of $\langle U(\infty, -\infty)\rangle_0$ as an additional factor. The factor $1/n!$ has been shown to be canceled when equivalent graphs are counted, and the contribution of such a disconnected graph is indeed just the product of the

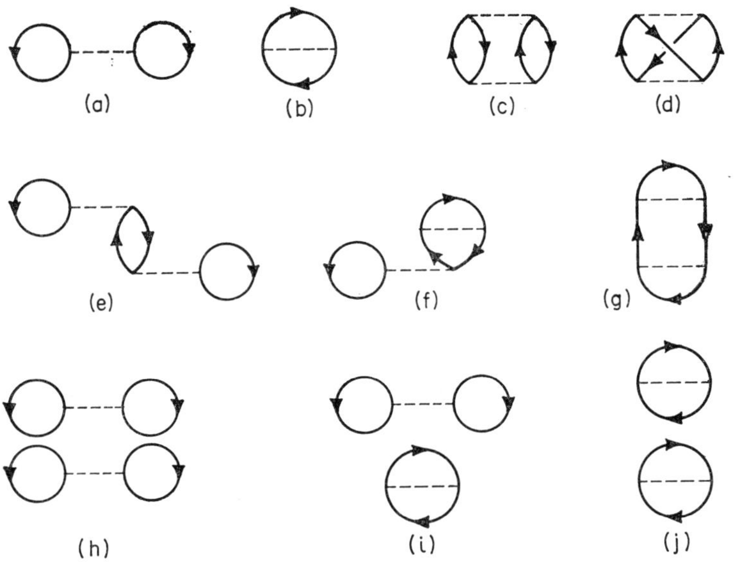

FIGURE 5 First and second-order graphs for $\langle U(\infty, -\infty)\rangle_0$

contribution of the linked part alone (the part connected to the endpoints) and a factor equal to the contribution to $\langle U(\infty, -\infty)\rangle_0$ of the disconnected part; thus the sum of all those graphs with the same linked part is just the product of $\langle U(\infty, -\infty)\rangle_0$ and the linked part alone. The result, then, is that the same expression $\langle U(\infty, -\infty)\rangle_0$ appears as a factor in the numerator, and is exactly canceled by the denominator. This cancellation is particularly important, since the dependence of the disconnected portions [and hence of $\langle U(\infty, -\infty)\rangle_0$ itself] on the volume V of the system, and on the time T during which the interaction is on, is rather nasty as V and T become infinite. In particular, a graph in $\langle U(\infty, -\infty)\rangle_0$ with p disconnected parts is proportional to $(VT)^p$, the sum of all graphs in $\langle U(\infty, -\infty)\rangle_0$ consisting of a single connected portion is of the form $-i(\Delta E)T$, and $\langle U(\infty, -\infty)\rangle_0$ itself turns out to be of the form $\exp[-i(\Delta E)T]$. Here $\Delta E/V$ is real and finite as V and T become infinite, and is in fact the shift in energy density of the true system with respect to the unperturbed system. This provides, incidentally, an alternative to Eq. (3.98) as a way of obtaining the true energy of the system.

To return to our propagator, however, we see now that our prescription may be applied to linked graphs only, and the denominator $\langle U(\infty, -\infty)\rangle_0$ ignored in compensation. Symbolically, we can write

$$G^c(k, t - t_0) = -i \sum^{(L)} (-1)^L \int \prod dt_i \sum_{\{k\}} \prod_i (-iv) \prod_r (iG_0^c). \quad (4.46)$$

Here the summation $\sum^{(L)}$ is over distinct linked graphs, the over-all factor $1/n!$ and the factors $\frac{1}{2}$ associated with vertices have been dropped as explained above, and the arguments of the interaction potential v and the propagators G_0^c are to be read off from the particular graph. The index i labels the vertices, and the index r labels the lines. The factor $(-1)^L$, where L is the number of closed fermion loops in the diagram, arises from the permutations of fermion operators associated with Wick's theorem (Eq. 4.36), as has been mentioned. It is to be recalled that for a line which begins and ends at the same vertex the time argument is to be taken as a negative infinitesimal $-\varepsilon$.

4.3 *Fourier transformed prescription*

This completes the expansion procedure for the propagator $G^c(k, t - t_0)$; it is particularly useful, however, to obtain the corresponding expansion for the time Fourier transform $G^c(k, \omega)$ (Eqs. 3.60, 68, 70, 74), for various reasons. First, the spectral function $\varrho(k, \omega)$ can be obtained directly from the imaginary part of $G^c(k, \omega)$; second, there occur frequently in the expansion of $G^c(k, t - t_0)$ integrations of the folding type, $\int_{-\infty}^{\infty} f(t - t_1)\, g(t_1)\, dt_1$, which become simple products on Fourier transformation.

We perform the Fourier transformation graph by graph. To begin with, we substitute for $G_0^c(k, t_i - t_j)$ its expression in terms of its Fourier transform (Eq. 3.59):

$$G_0^c(k, t_i - t_j) = (2\pi)^{-1} \int_{-\infty}^{\infty} G_0^c(k, \omega)\, e^{-i\omega(t_i - t_j)}\, d\omega, \quad (4.47)$$

where $G_0^c(k, \omega)$ is given by Eq. (3.85), and where an additional factor $e^{i\omega\varepsilon}$ should be inserted for a line that begins and ends at the same vertex, corresponding to the negative infinitesimal time argument mentioned previously, and here needed for such a line to make the ω integration unambiguous. When this substitution is made for every line, we find that the t-dependence is now entirely in the form of exponentials of the sort appearing in Eq. (4.47). We let each line of the graph be assigned a different integration variable ω_r, and since two lines enter and two leave each vertex, the t integration at that vertex will yield a factor of the general form

$$\int_{-\infty}^{\infty} \exp\left[i\left(\omega_r + \omega_s - \omega_{r'} - \omega_{s'}\right) t\right] dt = 2\pi\, \delta\left(\omega_r + \omega_s - \omega_{r'} - \omega_{s'}\right), \tag{4.48}$$

which can be interpreted as a sort of energy conservation in each virtual interaction. The ω_r's are independent variables, though, and not the free particle energies, so that this does not contradict the usual nonconservation of energy in virtual intermediate states. The two fixed vertices at t and t_0 will show a time dependence of the form $e^{-i\omega(t-t_0)}$, where the factors ω of t and t_0 are the same because of the energy conservation at all the internal vertices of the graph. Thus $G^c(k, t - t_0)$ is already cast into the form of Eq. (3.59), and we can read off the prescription for $G^c(k, \omega)$. There must be an over-all factor of 2π to compensate for the $(2\pi)^{-1}$ in Eq. (3.59), and factors $(2\pi)^{-1}$ at each line and 2π at each internal vertex, as seen above. Since, as we show below, the number of lines l is related to the number of internal vertices n by

$$l = 2n + 1, \tag{4.49}$$

the factors of 2π sort out to give $(2\pi)^{-n}$, or one factor $(2\pi)^{-1}$ for each vertex. The identical argument can be used to collect the factors i and $-i$ appearing in Eq. (4.46) to yield a factor $+i$ for each vertex.

(To get relation (4.49), we consider that each line has two ends, which are attached to vertices, so that if we count all of these ends we shall have counted each internal vertex four times, and the two external vertices once each. That is, $2l = 4n + 2$, from which our result follows.)

Our prescription is now this: we draw all possible distinct linked graphs with two external vertices, and arbitrarily many internal vertices. Each line is given its own momentum and energy variables k_r and ω_r, except that the lines attached to the external vertices must have the momentum and energy variables k and ω which appear as the arguments of $G^c(k, \omega)$. With each line is associated a factor $G_0^c(k_r, \omega_r)$, and with each vertex a factor $i\,(2\pi)^{-1} \times v_{rs,r's'}$. There is also an energy delta function $\delta(\omega_r + \omega_s - \omega_r' - \omega_s')$ associated with each vertex, though one of these has already been used, so to speak, to set the energy of one end line equal to that of the other. One can put this delta function back in if one lets ω and ω_0 be the energies of the two end lines, so that one obtains $G^c(k, \omega)\,\delta(\omega - \omega_0)$ from the prescription. In analogy to Eq. (4.46) we may write this symbolically thus:

$$G^c(k, \omega)\,\delta(\omega - \omega_0) = \sum^{(L)} (-1)^L \sum_{\{k_r\}} \int \prod d\omega_r$$
$$\times \prod_i [(i/2\pi)\, v(r, s; r', s')\, \delta(\omega_r + \omega_s - \omega_{r'} - \omega_{s'})] \prod_r [G_0^c(k_r, \omega_r)\, e^{i\omega_r\varepsilon}]. \tag{4.50}$$

(The factor $e^{i\omega_r\varepsilon}$ need only be retained, as has been mentioned, for lines which begin and end at the same vertex.)

This completes the prescription for calculating the single-particle propagator $G^c(k, \omega)$; the rules can be illustrated by writing down the contributions to the propagator of several of the graphs shown in Figures 2 and 3, namely, Figures 2(a), (b), (c), and 3(a), (c):

$$G^c_{2(a)}(k, \omega) = G_0^c(k, \omega) = [\omega - \varepsilon_k + i\eta\sigma(\bar{\varepsilon}_k)]^{-1}; \tag{4.51}$$

$$G^c_{2(b)}(k, \omega) = -(i/2\pi) \sum_{k_1} \int d\omega_1 G_0^c(k, \omega)\, v(k, k_1; k, k_1)\, G_0^c(k_1, \omega_1)$$
$$\times e^{i\omega_1\varepsilon}\, G_0^c(k, \omega); \tag{4.52}$$

$$G^c_{2(c)}(k, \omega) = (i/2\pi) \sum_{k_1} \int d\omega_1\, G_0^c(k, \omega)\, v(k, k_1; k_1, k)\, G_0^c(k_1, \omega_1)$$
$$\times e^{i\omega_1\varepsilon}\, G_0^c(k, \omega); \tag{4.53}$$

$$G^c_{3(a)}(k, \omega) = -(i/2\pi)^2 \sum_{k_1, k_2} \int d\omega_1\, d\omega_2\, G_0^c(k, \omega)\, v(k, k_1; k_2, k_3)$$
$$\times G_0^c(k_1, \omega_1)\, G_0^c(k_2, \omega_2)\, G_0^c(k_3, \omega_3)\, v(k_2, k_3; k, k_1)\, G_0^c(k, \omega)$$
$$(k_3 \equiv k + k_1 - k_2; \quad \omega_3 \equiv \omega + \omega_1 - \omega_2); \tag{4.54}$$

$$G^c_{3(c)}(k, \omega) = (i/2\pi)^2 \sum_{k_1, k_2} \int d\omega_1\, d\omega_2\, G_0^c(k, \omega)\, v(k, k_1; k, k_1)$$
$$\times G_0^c(k_1, \omega_1)\, e^{i\omega_1\varepsilon}\, G_0^c(k, \omega)\, v(k, k_2; k, k_2)$$
$$\times G_0^c(k_2, \omega_2)\, e^{i\omega_2\varepsilon}\, G_0^c(k, \omega). \tag{4.55}$$

4.4 *Hartree–Fock approximation; self-energy function*

In each of the three expressions (4.52, 53, 55), corresponding to Figures 2(b, c), 3(c), there appears the same integral over ω, which can be done very easily by using a contour integral procedure:

$$\int G_0^c(k_1, \omega_1)\, e^{i\omega_1 \varepsilon}\, d\omega_1 = \int_{-\infty}^{\infty} (\omega_1 - \varepsilon_1 \pm i\eta)^{-1}\, e^{i\omega_1 \varepsilon} d\omega_1$$

$$= 0, \quad \text{if } +i\eta, \text{ i.e., if } \varepsilon_1 > \mu;$$

$$= 2\pi i, \quad \text{if } -i\eta, \text{ i.e., if } \varepsilon_1 < \mu; \tag{4.56}$$

or, in short,

$$\int G_0^c(k_1, \omega_1)\, e^{i\omega_1 \varepsilon}\, d\omega_1 = 2\pi i \theta(-\bar{\varepsilon}_1). \tag{4.57}$$

Thus, the contribution of Figures 2(b) and 2(c) is just

$$G_{2(b)}^c + G_{2(c)}^c = \sum_{k_1} G_0^c(k, \omega)\, [v(k, k_1; k, k_1)$$

$$- v(k, k_1; k_1, k)]\, \theta(-\bar{\varepsilon}_1)\, G_0^c(k, \omega) \tag{4.58}$$

$$= G_0^c(k, \omega)\, V(k)\, G_0^c(k, \omega), \tag{4.59}$$

where $V(k)$ is the Hartree–Fock-type effective single-particle potential field due to the presence of all the other particles in free-particle (plane-wave) states. The contribution of Eq. (4.55) [Figure 3(c)], together with the corresponding exchange diagrams, also involves the same function $V(k)$, so that one has, finally, a whole family of graphs, extending to all orders in perturbation theory, but all consisting of the graphs of Figures 2(b, c) arranged end-to-end. This gives an approximation to $G^c(k, \omega)$ of the form

$$G_{HF}^c(k, \omega) = G_0^c(k, \omega) + G_0^c(k, \omega)\, V(k)\, G_0^c(k, \omega)$$

$$+ G_0^c(k, \omega)\, V(k)\, G_0^c(k, \omega)\, V(k)\, G_0^c(k, \omega) + \cdots. \tag{4.60}$$

This simple geometric series can be summed formally (ignoring the problem of convergence, which is tricky and probably irrelevant) to give

$$G_{HF}^c(k, \omega) = [\omega - \varepsilon_k - V(k) \pm i\eta]^{-1}. \tag{4.61}$$

It will be seen that this has a simple pole at $\omega = \varepsilon_k + V(k)$ corresponding to a single-particle energy appropriate to the effective single-particle potential $V(k)$; in this approximation there is still no damping of the single-particle excitation. The summing of the geometric series does not tell what to do with the infinitesimal $i\eta$ in the denominator; the requirement (3.74, 81, 29) that $G^c(k, \omega)$ consist of boundary values of the analytic propagator $G(k, z)$ demands a form analogous to (3.86):

$$G^c_{HF}(k, \omega) = [\omega - \varepsilon_k - V(k) + i\eta\sigma(\bar{\omega})]^{-1} \tag{4.62}$$

$$= [\omega - \varepsilon_k - V(k) + i\eta\sigma\{\bar{\varepsilon}_k + V(k)\}]^{-1}. \tag{4.63}$$

We see now an inconsistency between the form of $G^c(k, \omega)$ (Eq. 4.62) and the form of $V(k)$ (Eqs. 4.58, 59), since in $V(k)$ the states summed over are those for which $\varepsilon_k < \mu$, while the perturbed system has those states occupied for which $\varepsilon_k + V(k) < \mu$. That is, the perturbed system has a new value for the Fermi momentum, corresponding to the fact that if we take μ as given, then the particle density must be regarded as perturbed by the interaction. To make the approximation self-consistent it is necessary to modify the definition of $V(k)$ so that the sum is over those states for which $\varepsilon_k + V(k) < \mu$. This now gives exactly the Hartree–Fock approximation for the uniform medium (the perturbed wave-functions are still plane waves), and corresponds to using all the graphs of $G^c(k, \omega)$ in the internal line of Figures 2(b, c), since this would correspond to substituting $G^c(k_1, \omega_1)$ for $G^c_0(k_1, \omega_1)$ in Eqs. (4.52, 53, 55, 57), yielding, in Eq. (4.57), the result

$$\int G^c(k_1, \omega_1)\, e^{i\omega_1\varepsilon} d\omega_1 = 2\pi i\theta\,[\mu - \varepsilon_1 - V(k_1)], \tag{4.64}$$

and hence

$$V(k) = \sum_{k_1} [v(k, k_1; k, k_1) - v(k, k_1; k_1, k)]\, \theta\,[\mu - \varepsilon_1 - V(k_1)]. \tag{4.65}$$

It is true that this introduction of self-consistency here makes only a minor modification, because the wave functions are necessarily plane waves; however, the same procedure yields the Hartree-Fock approximation also in the nontrivial case where the system is not translation-invariant and the wave functions are therefore not plane waves.

We should try now to relate the form of the graphs in Figures 2(b, c), to the physical significance of the corresponding contributions to the propagator. The rule of using an infinitesimal time ε for a line that begins and ends at the same vertex signifies that the tail of such a line is to be taken as if it were at a slightly later time than the head; this corresponds to the propagation of a hole, with momentum $-k_1$ (less than the Fermi momentum) between these two times. This hole does not itself play any rôle—the only significance of this picture is to represent the absorption and immediate re-emission of a normally unexcited particle of momentum k_1. There are, of course, a very large number of such unexcited particles, and rather than indicating the presence of all of them by lines in our picture, we represent the absorption of one of them by the appearance of a hole, and its reappearance by the vanishing of the hole. Thus, we see that Figures 2(b, c) represent pictorically the virtual direct and exchange scattering of the particle we are interested in with each of the unexcited particles already present in the unperturbed system.

The spectral function corresponding to this approximation for $G^c(k, \omega)$ can be obtained in principle from Eq. (3.70), but since $G^c(k, \omega)$ is no different in form from $G_0^c(k, \omega)$, it is easy to see that the spectral function is simply a displaced delta function:

$$\varrho_{HF}(k, \omega) = \delta\,[\omega - \varepsilon_k - V(k)]. \tag{4.66}$$

One can use this form for $\varrho\,(k, \omega)$ in Eqs. (3.92, 97) to obtain the corresponding approximations to the total particle number (as a function of μ) and energy. Since $f^-(\bar{\omega})$ reduces to the simple step function $\theta(-\bar{\omega})$ for zero temperature, we get

$$\mathcal{N} = \sum_k \theta\,[-\bar{\varepsilon}_k - V(k)]; \tag{4.67}$$

$$\begin{aligned} E &= \tfrac{1}{2} \sum_k \theta\,[-\bar{\varepsilon}_k - V(k)]\,[\varepsilon_k + V(k) + \varepsilon_k] \\ &= \sum_k \theta\,[-\bar{\varepsilon}_k - V(k)]\,[\varepsilon_k + \tfrac{1}{2}V(k)], \end{aligned} \tag{4.68}$$

the proper Hartree-Fock results (see discussion following Eq. (3.98).

Quite generally, one can see that every graph contributing to $G^c(k, \omega)$ consists of one or more subgraphs joined to each other by just a single line; these subgraphs (if they cannot be further subdivided in the same way) are called "self-energy graphs" (or more correctly, "proper self-energy graphs"). Because of energy-momentum conservation, the single lines connecting the self-energy subgraphs will each correspond to a factor $G_0^c(k, \omega)$ with the same fixed arguments k and ω. Since all possible self-energy subgraphs will appear in the different graphs of $G^c(k, \omega)$, we can let $M^c(k, \omega)$ represent the sum of all possible self-energy subgraphs, and write

$$\begin{aligned} G^c(k, \omega) &= G_0^c(k, \omega) + G_0^c(k, \omega)\,M^c(k, \omega)\,G_0^c(k, \omega) \\ &\quad + G_0^c(k, \omega)\,M^c(k, \omega)\,G_0^c(k, \omega)\,M^c(k, \omega)\,G_0^c(k, \omega) + \cdots. \end{aligned} \tag{4.69}$$

This is again a simple geometric series, like that in Eq. (4.60), and can also be formally summed, yielding

$$G^c(k, \omega) = [\omega - \varepsilon_k - M^c(k, \omega)]^{-1}. \tag{4.70}$$

The term $i\eta$ in the denominator no longer causes any difficulty, since $M^c(k, \omega)$ is in general complex, and can in fact be shown to have an imaginary part which is positive for $\omega < \mu$ and negative for $\omega > \mu$. (This is implicit in Eq. (3.70) together with the fact that $\varrho(k, \omega)$ is positive for FD statistics.)

We see now that we need only calculate the self-energy function $M^c(k, \omega)$, and that a perturbative approximation to $M^c(k, \omega)$ yields, through Eq. (4.70), a nonperturbative approximation to $G^c(k, \omega)$. This alone does not guarantee that this is better in every way than a perturbative approximation to $G^c(k, \omega)$, but it does allow $G^c(k, \omega)$ to depart from the simple pole behavior of $G_0^c(k, \omega)$, and thus display finite lifetime effects. The spectral function $\varrho(k, \omega)$ can be obtained directly from the imaginary part of $G^c(k, \omega)$, using Eq. (3.70), and will consist of a delta function if $M^c(k, \omega)$ is real, but have a nonzero width if $M^c(k, \omega)$ is complex.

One can also go beyond the perturbation expansion in the calculation of $M^c(k, \omega)$ by noting that the same graphs which make up $G^c(k, \omega)$ itself also appear as subgraphs of $M^c(k, \omega)$. Each irreducible graph of $M^c(k, \omega)$, that is, one which contains no self-energy subgraphs, can be taken to represent the infinite class of graphs obtained by replacing $G_0^c(k, \omega)$ by $G^c(k, \omega)$ wherever it appears. Of course, $G^c(k, \omega)$ is not known explicitly, but the resulting equation, expressing $M^c(k, \omega)$ in terms of $G^c(k, \omega)$, can in principle be solved in connection with Eq. (4.70), which expresses $G^c(k, \omega)$ in terms of $M^c(k, \omega)$. This is, in fact, exactly what we did in the case of the graphs in Figures 2(b, c) to obtain the self-consistent Hartree–Fock approximation.

4.5 *Large-volume limit*

We have mentioned previously (Eqs. 3.94, 95, 98) what happens to the total particle number and energy on passing to the limit of large volumes. We should now examine in the same limit our prescription for calculating the propagator and see in particular that the spectral function $\varrho(k, \omega)$ which appears in the calculation of $\mathcal{N}$ and E, has at least formally a finite limit as $V \to \infty$, as was assumed in finding Eqs. (3.95, 98).

In calculating the contribution of a graph, the volume dependence enters not only through the transition from sums to integrals (Eq. 3.94), but also through the matrix elements of the two-body interaction v, which (Eq. 2.70) takes the limiting form, in the case of a local potential,

$$v(1, 2; 1', 2') = (2\pi)^3 V^{-1} v(k_1 - k_1') \, \delta(k_1 + k_2, k_1' + k_2'). \qquad (4.71)$$

In an nth-order graph there are n Kronecker deltas of the type appearing above; $n - 1$ of these serve to reduce the number of k-summations associated with the lines of the diagram, while the last one appears as an overall factor like the one in Eq. (3.31). If l is the number of lines, then since the two end lines already have their momentum fixed, and the number of further restrictions is $n - 1$, the number of independent k-summations remaining is $(l - 2) - (n - 1)$, which by Eq. (4.49), is just n. The factors $V/(2\pi)^3$ resulting from the conversion of these sums to integrals (Eq. 3.94) just compensate the factors $(2\pi)^3/V$ associated (Eq. 4.71) with the interaction matrix elements, and so we find that to all orders $G^c(k, \omega)$ has a finite limit as $V \to \infty$. The spectral function also, which can be obtained directly from the imaginary part of $G^c(k, \omega)$ (Eq. 3.70), has thus a finite limit, at least formally, so that the limiting procedures for $\mathcal{N}$ and E are formally justified.

4.6 The many-boson system

In its unperturbed ground state (Eq. 2.27), the many-boson system has its zero-momentum state (single-particle state of lowest energy) macroscopically occupied. This is to say that in the large-volume limit, the particle density being held fixed, the number of particles in the zero-momentum state is proportional to the volume. This is a roundabout way of stating the obvious facts that all the particles are in the zero-momentum state, and that the total number of particles is proportional to the volume; this is in contrast, though, to the case of fermions, or of interacting bosons with no condensed phase, since in both of these cases, the occupation number of each single-particle state remains fixed as $V \to \infty$, while the density of states) i.e. the number of single-particle states per unit energy range), which increases proportionally to V, provides the increase in particle number.

If the boson system we are considering has no condensed phase in its true ground state, then we probably cannot apply a perturbation expansion at all, in view of the qualitative difference from the unperturbed system; the only examples in nature would be solids. We are thus led to restrict our attention here to the system with a condensed phase in its true ground state, that is, with a large number of particles in the zero-momentum state. The rule for handling this case is very simple, both at $T = 0$ and at finite temperatures, but is somewhat more difficult to justify rigorously. The ground-state case has been treated by Beliaev [39], and by Kromminga and Bolsterli [41], while the finite-temperature case is amenable to similar treatment.

This method is based on the fact that finite changes in the number of zero-momentum particles produce essentially negligible changes in the system; in consequence, one can treat the occupation number of this state as a constant, while at the same time introducing an artificial sink and source for particles in other states, with probability amplitudes equal to those for scattering into and out of the zero-momentum state. To start with, we work with a modified Hamiltonian $\bar{H}$:

$$\bar{H} = H - \mu \mathbf{N}, \tag{4.72}$$

which, since N is fixed, provides the same physical description of the system, but normalizes each single-particle energy from ε_k to $\bar{\varepsilon}_k$ (Eq. 3.21). The purpose of this modification is to enable us to adjust μ so that a change in the number of zero-momentum particles will not change the energy of the system. We now replace the zero-momentum-state operator a_0 by the c-numbers $N_0^{1/2}$, both in $\bar{H}$ and in any propagator we may wish to calculate.* This gives a further modified Hamiltonian $\bar{H}(\mu, N_0)$, involving the two parameters μ and N_0, and no zero-momentum operators; if these parameters are chosen correctly, then the propagators calculated with this Hamiltonian (without particle-number restriction) will be equal, to dominant order in the volume V, to the true propagators calculated with $\bar{H}$ (particle number fixed). These in turn differ only by factors of the form $e^{\pm i\mu t}$ from those calculated with H and discussed in Section 3. For two-body forces (2.68), $\bar{H}(\mu, N_0)$ takes the following form:

$$\begin{aligned}
\bar{H}(\mu, N_0) = &-\mu N_0 + \sum{}' \bar{\varepsilon}_k a_k^\dagger a_k + \tfrac{1}{2} v(0, 0; 0, 0) N_0^2 \\
&+ \tfrac{1}{2} N_0 \sum{}' v(1, 2; 0, 0)\, a_1^\dagger a_2^\dagger + \tfrac{1}{2} N_0 \sum{}' v(0, 0; 1', 2')\, a_{1'} a_{2'} \\
&+ N_0 \sum{}' [v(1, 0; 1, 0) + v(1, 0; 0, 1)]\, a_1^\dagger a_1 \\
&+ N_0^{1/2} \sum{}' v(1, 2; 1', 0)\, a_1^\dagger a_2^\dagger a_{1'} + N_0^{1/2} \sum{}' v(1, 0; 1', 2')\, a_1^\dagger a_{1'} a_{2'} \\
&+ \tfrac{1}{2} \sum{}' v(1, 2; 1', 2')\, a_1^\dagger a_2^\dagger a_{1'} a_{2'},
\end{aligned} \tag{4.73}$$

where the summations Σ' are over all momenta other than zero.

The total particle number can be calculated from Eq. (3.90), but the corresponding expression (3.96) for the energy cannot be used, because the interaction terms in (4.73) are no longer all of the two-body form (2.68). The

---

* One argument for doing this (see ref. 40) says that the commutator of a_0 and $a_0^\dagger$, which is unity, is so small compared to their product $\mathbf{n}_0$, whose expectation value is proportional to V, that they may be treated as commuting operators. Since they also commute with all other operators a_k and $a_k^\dagger$, they may be replaced by c-numbers. This argument does not take into account the exponential dependence on V of the numerator and denominator of the expression (4.35) for $G^c(k, t - t_0)$, and the necessity for strict cancellation, but clearly represents a true insight.

ground-state expectation value of (4.73) must be calculated for an arbitrary N_0, either by an extension of the perturbation technique to expectation values of products of three or four operators, or else by a modification of (3.96) appropriate to the form of (4.73):

$$\bar{E}(\mu, N_0) = -\mu N_0 + \tfrac{1}{2}v(0,0;0,0)\,N_0^2 + N_0^2 \int_{N_0}^{\infty} (N_0')^{-3}\,dN_0'\,\mathrm{Re}\sum{}' \langle \mathbf{a}_k^\dagger(0)\,[i\dot{\mathbf{a}}_k(0) + \bar{\varepsilon}_k \mathbf{a}_k(0)]\rangle_{\mu, N_0'} \tag{4.74}$$

This can be justified by comparing $\bar{E}$ and $\partial\bar{E}/\partial N_0$ (which is equal to $\langle \partial \bar{H}/\partial N_0\rangle$ because of the stationary property of $\langle \bar{H}\rangle$) with the expectation value appearing on the right.

The parameters μ and N_0 are determined by the conditions

$$\partial\bar{E}(\mu, N_0)/\partial N_0 = 0, \tag{4.75}$$

and

$$\mathcal{N}(\mu, N_0) = N, \tag{4.76}$$

where N is the given total particle number. The condition (4.75) expresses the fact that μ is chosen to make $\bar{E}$ insensitive to changes in N_0.

It is seen that (4.73) is formally a Hamiltonian for a system with particle number not conserved, and a variety of virtual processes possible. The first three terms can be taken as the unperturbed Hamiltonian H_0 (though the constant terms play no active rôle here), and the remaining terms taken as H'. The unperturbed ground state is simply that in which no particles (of non-zero momentum) are present, i.e., a vacuum-like state, while the perturbing terms allow particle production and annihilation. In these circumstances we can apply Wick's theorem, as in the preceding development, and work out the graphical expansion procedure much as before. We now have seven different types of vertex, shown in Figure 6, corresponding to the seven interaction terms in Eq. (4.73), which in turn correspond to the different ways in which zero-momentum particles can participate in a virtual scattering pro-

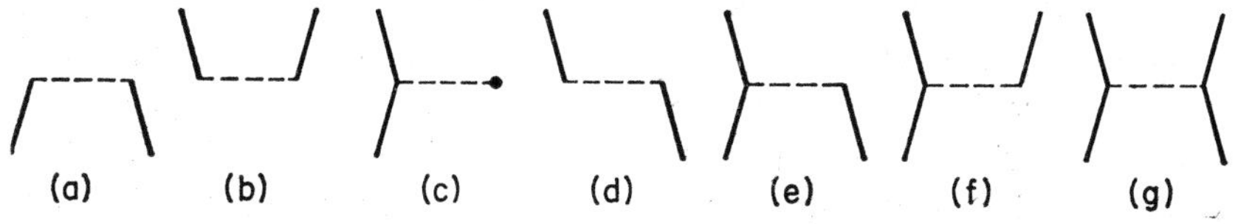

FIGURE 6 Vertex parts for the many-boson problem

cess. It is not hard to check that in the large-volume limit all of these vertex parts give contributions to a propagator which are finite in the limit, the extra factors N_0 and $N_0^{1/2}$ compensating for the smaller number of k-summations, each of which, it will be remembered (Eq. 3.94), introduces a factor proportional to V.

The use of the modified Hamiltonian $\bar{H}$ (Eq. 4.72) is equivalent, as has been mentioned, to displacing each single-particle energy by μ, that is, replacing each ε_k by $\bar{\varepsilon}_k$. In consequence, the propagator $G^c(k, t)$ which we calculate differs from the previous definition (Eq. 3.34) by an additional factor $e^{i\mu t}$. If we use $\bar{\omega}$ in taking the Fourier transform, this factor is reabsorbed and we obtain the same propagator $G^c(k, \omega)$ defined before (3.60, 62):

$$G^c(k, \omega) = \int G^c(k, t)\, e^{i\bar{\omega}t}\, dt. \tag{4.77}$$

However, in the Fourier transformed Feynman graph prescription it is $\sum \bar{\omega}$ which is conserved at each vertex; this is equivalent to the ordinary conservation of the energy $\sum \omega$ if in the latter case an energy μ is assigned to each missing zero-momentum particle.

The final prescription for $G^c(k, \omega)$ is as follows: for $k = 0$, we find

$$G^c(0, t) = -iN_0. \tag{4.78}$$

The Fourier transform is then (Eq. 4.77):

$$G^c(0, \omega) = -2\pi i N_0 \delta(\bar{\omega}). \tag{4.79}$$

For $k \neq 0$, we draw all possible linked graphs, using the vertex parts of Figure 6 (only graphs which allow all arrows to point forward in time will be nonzero, though); each line r is assigned a momentum k_r and an energy variable ω_r, except that the lines attached to the external vertices have the given momentum k, and the given energies ω and ω_0. For each line there is a factor $i\,(2\pi)^{-1}\, G_0^c\,(k_r, \omega_r)$, with

$$G_0^c(k, \omega) = (\omega - \varepsilon_k + i\eta)^{-1}, \tag{4.80}$$

a summation over k_r, and an integration over ω_r. For each vertex there is a factor $(-2\pi i)$, an appropriate matrix element $v_{(r,s;r',s')}$, a factor $N_0^{1/2}$ for each absent (zero-momentum) line and an energy delta function $\delta\,(\omega_r + \omega_s - \omega_{r'} - \omega_{s'})$. One or more of the momenta $k_r, k_s, k_{r'}, k_{s'}$ may be zero, in which case the corresponding energy variable ω is to be set equal to μ. (The factors $\frac{1}{2}$ appearing in some of the terms of Eq. (4.73) go out as before, if only distinct nonidentical graphs are included.) An overall factor $(-2\pi i)$ completes the prescription. This calculation yields an expression with the overall

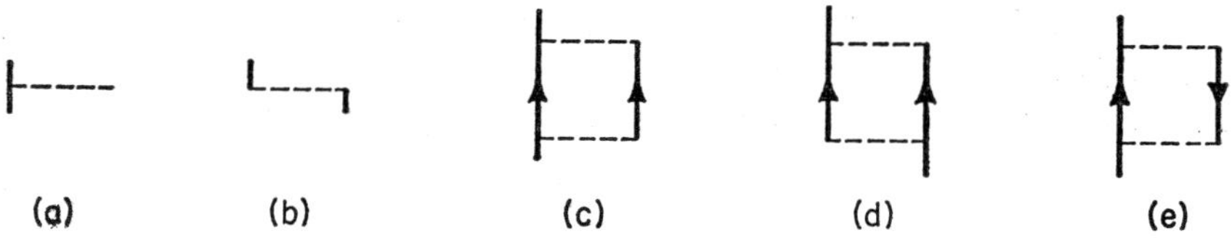

FIGURE 7 Proper self-energy graphs for the many-boson system

factor $\delta(\omega - \omega_0)$, as in Eq. (4.50), and thus yields a prescription for $G^c(k, \omega)\,\delta(\omega - \omega_0)$.

The discussion of self-energy graphs in connection with the fermion propagator carries through here, and so we can give as examples the contributions to $M^c(k, \omega)$ due to the self-energy graphs shown in Figure 7:

$$M^c_{7(a)}(k, \omega) = N_0 v\,(k, 0; k, 0); \tag{4.81}$$

$$M^c_{7(b)}(k, \omega) = N_0 v\,(k, 0; 0, k); \tag{4.82}$$

$$M^c_{7(c)}(k, \omega) = i\,(2\pi)^{-1} N_0 \sum_{k_1}{}' \int d\omega_1 v\,(k, 0; k_1, k - k_1) \times v\,(k_1, k - k_1; k, 0)\, G^c_0(k_1, \omega_1)\, G^c_0(k - k_1, \omega - \bar{\omega}_1); \tag{4.83}$$

$$M^c_{7(d)}(k, \omega) = i\,(2\pi)^{-1} N_0 \sum_{k_1}{}' \int d\omega_1 v\,(k, 0; k_1, k - k_1) \times v\,(k_1, k - k_1; 0, k)\, G^c_0(k_1, \omega_1)\, G^c_0(k - k_1, \omega - \bar{\omega}_1); \tag{4.84}$$

$$M^c_{7(e)}(k, \omega) = i\,(2\pi)^{-1} N_0 \sum_{k_1}{}' \int d\omega_1 v\,(k, k_1; k + k_1, 0) \times v\,(k + k_1, 0; k, k_1)\, G^c_0(k_1, \omega_1)\, G^c_0(k + k_1, \omega + \bar{\omega}_1) = 0. \tag{4.85}$$

The vanishing of $M^c_{7(e)}$ can be shown by doing the ω_1 integration. (The path of integration can be closed in the upper half plane, where the integrand is analytic), or by noting, in terms of time variables, that $G^c_0(k, t)$ vanishes for negative t, so that the function $M^c_{7(e)}(k, t)$ vanishes no matter which vertex is earlier in time.

It is interesting to note that a simple, completely soluble theory, corresponding to the Bogoliubov canonical transformation [18] method, is obtained by keeping only the vertices of Figures 6(a–d).

For the ground state problem, one further relation, similar to Eq. (4.75), can be obtained, namely, that

$$\begin{aligned} \partial \bar{E}\,(\mu, N_0)/\partial\mu &= \partial\,\langle \bar{H} \rangle/\partial\mu \\ &= \langle \partial \bar{H}/\partial\mu \rangle \\ &= -N_0 - \langle \mathbf{N}' \rangle \\ &= -N, \end{aligned} \tag{4.86}$$

provided μ and N_0 are set equal to their proper values. From this and Eq.(4.75) we can see that the total differential of $\bar{E}$ (N_0 being held equal to its correct value $N_0(\mu)$, and T and V being held fixed) is given by

$$d\bar{E} = -N\,d\mu, \tag{4.87}$$

so that that of E is given by

$$\begin{aligned} dE &= d\bar{E} + \mu\,dN + N\,d\mu \\ &= \mu\,dN. \end{aligned} \tag{4.88}$$

But E in fact depends only on N, since μ and N_0 are determined by our choice of N, so that μ is identical to the separation energy, as anticipated by the notation:

$$\partial E\,(N, T, V)/\partial N = \mu \quad (T = 0). \tag{4.89}$$

In fact, since μ is the zero-temperature limit of the Gibbs function per particle, it follows, from the thermodynamic relation

$$G = E - TS + PV \tag{4.90}$$

that

$$\bar{E} \equiv -PV, \tag{4.91}$$

where P is the pressure.

This completes for the present the discussion of the ground state expansion. It is clear, and will be seen in the next section, that the ground state problem can be treated as a special case of the finite temperature problem; further discussion is perhaps best left until the more general theory has been developed.

5

Perturbation expansion—Finite temperature

IT WILL be recalled that the basic elements of the ground state expansion were the perturbation expansion of the time-development operator $U(t, t_0)$, (Eqs. 4.2, 13, 27), the use of the adiabatic theorem to obtain expressions (Eqs. 4.18, 19) for the true ground state vector, and Wick's theorem (Eq. 4.36), which permits one to reduce each term of the perturbation expansion to a sum of products of simple factors. It turns out that where the grand ensemble can be used, the finite-temperature problem can be handled by a remarkably similar method, which has the conceptual advantage (especially for an infinite system) of not making use of the adiabatic theorem. There are two basic mathematical facts which are responsible for this close similarity: first, that the grand partition function and the statistical expectation values with which we work (Eqs. 3.3) can be expressed in terms of the same time-development operator $U(t, t_0)$, analytically continued to complex values of the time arguments; and second, that statistical expectation values satisfy a modified form of Wick's theorem (the "thermodynamic Wick's theorem") [26, 54, 30] identical in form to Eq. (4.36), but with ground-state expectation values replaced by statistical averages. The boson system with a condensed phase must be treated basically by means of the canonical ensemble, for which this discussion does not apply; however, as we have mentioned, it can be reduced to an equivalent problem in which the macroscopically occupied state does not play a dynamical rôle, and this equivalent problem can be treated in the grand ensemble, using the methods we are describing.

We shall find that in the expressions analogous to Eqs. (4.34, 35) for example, the limits of integration over the variables t become τ, an arbitrary complex time, and $\tau - i\beta$. The integrations can be performed over a fairly arbitrary contour [37], between these two points; the contour must, however, pass through the two points t_0 and t which appear as arguments in the propagator. The theory can be developed for an arbitrary contour, as long as one is working directly with time variables, but only for certain special choices can the development in terms of Fourier transforms, analogous to Eq. (4.47), be worked out.

One of these choices [26, 30, 32] is the contour running along the imaginary axis from 0 to $-i\beta$; since this is a finite domain, we get a Fourier sum, and the Fourier transform consists of a discrete set of values, which are in fact equal to the analytic propagator $G(k, z)$ (Eq. 3.80), evaluated at a discrete set of points in the complex energy plane [36]. An alternative choice is a contour which is distorted to include the entire real axis. Under appropriate circumstances, the contribution of portions of the contour off the real axis can be made negligible, so that the real-time Fourier transform can be taken, and functions of a real energy argument be calculated directly without analytic continuation. A third expansion procedure [55, 56] results from performing the time integrations explicitly for each possible succession of intermediate states. The resulting expressions bear a close resemblance to those of conventional time-independent perturbation theory, with energy denominators associated with the sums over intermediate states, but with the additional feature of statistical weighting factors in appropriate places. We shall sketch briefly the derivation and uses of these alternative methods.

5.1 The analytic propagator $G(k, t - t_0)$

To begin with, then, we look at the expression (Eq. 3.6) for the grand partition function $\mathcal{Q}$, and notice that it can be expressed as a statistical average in the interaction representation, that is, using the unperturbed energies in the weighting factor. Thus (Eqs. 3.6, 7, 14, 15),

$$\begin{aligned}\mathcal{Q} &= \mathrm{Tr}\, e^{-\beta\bar{H}} \\ &= \mathrm{Tr}\,[e^{\beta\mu N}\, e^{-\beta H}] \\ &= \mathrm{Tr}\,[e^{-\beta\bar{H}_0}\, e^{\beta H_0}\, e^{-\beta H}] \\ &= \mathcal{Q}_0 \,\langle e^{\beta H_0}\, e^{-\beta H}\rangle_0 \\ &= \mathcal{Q}_0 \,\langle U(-i\beta)\rangle_0, \end{aligned} \tag{5.1}$$

where the use of the imaginary argument in $U(t)$ (Eq. 4.2) is here quite valid. In general we must be careful, in using complex arguments, that all expectation values which we use are in fact finite; this will impose certain restrictions on the values of complex arguments, which we shall determine later in this chapter. The cyclic property of the trace permits a useful generalization of Eq. (5.1) above: We could have written

$$\begin{aligned}\mathcal{Q} &= \mathrm{Tr}\,[e^{-\beta \bar{H}_0}\, e^{(\beta+i\tau)H_0}\, e^{-(\beta+i\tau)H}\, e^{i\tau H}\, e^{-i\tau H_0}] \\ &= \mathcal{Q}_0 \langle U(\tau - i\beta)\, U^{\dagger}(\tau)\rangle_0 . \end{aligned} \tag{5.2}$$

Here τ is an arbitrary complex number, so that we must be careful in defining $U^{\dagger}(\tau)$; it denotes here, not the hermitian conjugate, but rather what one might call the "hermitian adjoint function":

$$U^{\dagger}(\tau) = [U(\tau^*)]^{\dagger} = e^{\,i\tau H} e^{-i\tau H_0}. \tag{5.3}$$

Note that $U(\tau)$ is not unitary if τ is complex, but does satisfy

$$U(\tau)\, U^{\dagger}(\tau) = 1 . \tag{5.4}$$

We use the combination appearing in Eq. (5.2) to define the analytic continuation of $U(t, t_0)$: using the definition (Eq. 4.13) for arbitrary complex t and t_0,

$$U(t, t_0) = U(t)\, U^{\dagger}(t_0). \tag{5.5}$$

Thus we can write, for arbitrary complex τ,

$$\mathcal{Q}/\mathcal{Q}_0 = \langle U(\tau - i\beta, \tau)\rangle_0 . \tag{5.6}$$

We find that our propagators, too, can be expressed in terms of U. The basic propagators $G^{\pm}$ (Eq. 3.12, 13) are of the general form (see also Eq. (4.21)

$$G_{AB}(t, t_0) = \langle \mathbf{A}(t)\, \mathbf{B}(t_0)\rangle = \mathcal{Q}^{-1}\, \mathrm{Tr}\,[e^{-\beta \bar{H}}\, \mathbf{A}(t)\, \mathbf{B}(t_0)], \tag{5.7}$$

where the Heisenberg operators are defined by Eq. (2.81), and the true statistical average (using $e^{-\beta \bar{H}}$ as the weighting operator) is indicated. The times t and t_0 were originally taken as real, but the definition (Eq. 2.81) can be continued to complex values provided the sums over states implied by Eq. (5.7) converge. We proceed, in analogy to Eq. (4.22), by an argument similar to that just used for $\mathcal{Q}$, with the additional feature that we express the Heisenberg operators $\mathbf{A}(t)$ and $\mathbf{B}(t)$ in terms of the interaction representation operators by the use of Eq. (4.1), where again the formal extension to complex time arguments is trivial:

$$G_{AB}(t, t_0) = \mathcal{Q}^{-1} \operatorname{Tr} [e^{-\beta \bar{H}} \mathbf{A}(t) \mathbf{B}(t_0)]$$

$$= \mathcal{Q}^{-1} \operatorname{Tr} [e^{-\beta \bar{H}_0} e^{(\beta + i\tau)H_0} e^{-(\beta+i\tau)H} \mathbf{A}(t) \mathbf{B}(t_0) e^{i\tau H} e^{-i\tau H_0}]$$

$$= \mathcal{Q}^{-1} \operatorname{Tr} [e^{-\beta \bar{H}_0} U(\tau - i\beta) U^\dagger(t) A(t) U(t)$$

$$\times U^\dagger(t_0) B(t_0) U(t_0) U^\dagger(\tau)]$$

$$= (\mathcal{Q}/\mathcal{Q}_0)^{-1} \langle U(\tau - i\beta, t) A(t) U(t, t_0) B(t_0) U(t_0, \tau)\rangle_0. \qquad (5.8)$$

(The equivalence of the first and second lines follows from the cancellation of all the factors involving H_0, using the cyclic property of the trace, and then the cancellation of the factors $e^{\pm i\tau H}$; a factor $e^{\beta\mu N}$, which commutes with all the other exponential factors, remains in both lines.) This expression for the propagator, together with Eq. (5.6) for the grand partition function, is to be compared with Eq. (4.22), which has the same form, apart from the replacement of $\pm\infty$ by the complex times $\tau - i\beta$ and τ, and the absence of any adiabatic limiting procedure. Such a limiting procedure is to be compared with that of taking the limit $\beta \to \infty$, which should also give the ground state properties. We shall mention this again in an appropriate place.

The domain of analyticity of $G_{AB}(t, t_0)$ can be obtained on the assumptions that the exponentials $e^{-\beta \bar{H}}$ and $e^{\pm iHt}$ dominate in any sums over states, and that the operators A and B have nonzero matrix elements only between states whose particle numbers differ by at most a finite number (e.g., by unity, if A and B are simple annihilation or creation operators). If the operator A raises the particle number by δn, say (which would be ± 1 for creation and annihilation operators respectively), then we can replace H by $\bar{H}$ in the expression (Eq. 4.1) for $\mathbf{A}(t)$ by making the following change:

$$\mathbf{A}(t) = e^{iHt} A e^{-iHt}$$

$$= e^{i\bar{H}t} e^{i\mu N t} A e^{-i\mu N t} e^{-i\bar{H}t}$$

$$= e^{i\mu\delta n t} e^{i\bar{H}t} A e^{-i\bar{H}t}, \qquad (5.9)$$

since an operator $\mathbf{N}$ to the left of A always takes on a value greater by δn than one to the right. (The above discussion is trivial if particle number is not conserved, since in that case $\mu = 0$.) The extra factor $e^{i\mu\delta n t}$ is of course an entire function of t, and does not restrict the domain of analyticity. The operator B must lower the particle number by the same amount δn for $G_{AB}(t, t_0)$ not to vanish, so we can take out a factor $e^{i\mu\delta n(t-t_0)}$, and obtain

$$G_{AB}(t, t_0) = \mathcal{Q}^{-1} e^{i\mu\delta n(t-t_0)} \operatorname{Tr} [e^{-\beta\bar{H}} e^{i\bar{H}t} A\, e^{-i\bar{H}(t-t_0)} B\, e^{-i\bar{H}t_0}]. \qquad (5.10)$$

The advantage of this expression is that the time dependence is explicit, and apart from the innocuous factor $\exp [i\mu\delta n\,(t - t_0)]$ involves only the single operator $\bar{H}$, whose spectrum is bounded below and unbounded above. Using the cyclic property of the trace, and expanding in eigenstates of $\bar{H}$, we get

$$G_{AB}(t, t_0) = \mathcal{Q}^{-1} e^{i\mu\delta n(t-t_0)} \operatorname{Tr} [e^{i\bar{H}(t-t_0+i\beta)} A\, e^{-i\bar{H}(t-t_0)} B] \qquad (5.11)$$

$$= \mathcal{Q}^{-1} e^{i\mu\delta n(t-t_0)} \sum_{\lambda,\nu} e^{i\bar{E}_\lambda(t-t_0+i\beta)} \langle\lambda\,|A|\,\nu\rangle\, e^{-i\bar{E}_\nu(t-t_0)} \langle\nu\,|B|\,\lambda\rangle, \qquad (5.12)$$

where λ and ν label the eigenstates of $\bar{H}$ (Eq. 3.9). We see already that G_{AB} is a function of the difference $t - t_0$, as before, and on the assumption that the exponentials dominate the convergence of the sums over λ and ν, we see that they converge exponentially whenever

$$-\beta < \operatorname{Im}(t - t_0) < 0, \qquad (5.13)$$

and so $G_{AB}(t - t_0)$ is analytic in this strip. The domain of analyticity may of course extend beyond this, and indeed the unperturbed propagators $G_0^{\pm}(k, t - t_0)$, which are of the same form, are entire functions of $t - t_0$ (Eqs. 3.17, 18); in general, however, the propagator is analytic only in this strip. The limit of an analytic function on the boundary of its domain of analyticity is a generalized function, or distribution (e.g., the delta-function and its derivatives) [49], so that $G_{AB}(t - t_0)$ may be regarded as well defined for

$$-\beta \leqq \operatorname{Im}(t - t_0) \leqslant 0. \qquad (5.14)$$

This result is confirmed by the expressions (Eq. 3.56) for the propagators $G^{\pm}(k, t)$, since the functions $f^{\pm}(\bar{\omega})$ behave like $e^{-\beta|\bar{\omega}|}$ as $\omega \to \pm\infty$ for $f^{+}(\bar{\omega})$ and $f^{-}(\bar{\omega})$ respectively. Since the spectral function $\varrho\,(k, \omega)$ does not blow up exponentially as $\omega \to \pm\infty$ (its integral is unity), we see that $G^{+}(k, t)$, which is of the form of $G_{AB}(t)$, is well defined for $-\beta \leqslant \operatorname{Im} t \leqslant 0$, and analytic on the interior of this domain , while $G^{-}(k, t)$, which has the form of $G_{AB}(-t)$, is well defined for $0 \leqslant \operatorname{Im} t \leqslant \beta$.

The procedure we now follow is exactly analogous to that for the ground-state problem. The differential equation (Eq. 4.15), satisfied by $U(t, t_0)$ for real t and t_0, is satisfied by the analytically continued function as well, provided we retain the definition (Eq. 4.6) for the interaction Hamiltonian $H'(t)$

for complex t. The boundary condition (4.16) for $U(t, t_0)$ is still satisfied, and so, therefore, is the integral equation (Eq. 4.23), except that we must now pick a contour C running from t_0 to t in the complex plane. The iterated solution (Eq. 4.24) still holds, and it is here convenient (though not necessary) to let all of the integrations lie along the same contour C, or portions of it. The argument leading to Eq. (4.27) carries over, provided we generalize the time-ordering operator T to an operator T_C which orders factors along the contour C [37]. That is, it prescribes that the operators it is applied to be rearranged in the order in which their arguments lie along C, with those nearest the end at t_0 to the right, and those nearest the end at t to the left. The operator T is of course a special case of T_C. So we can write, for arbitrary complex t and t_0,

$$U(t, t_0) = \sum_{n=0}^{\infty} (-i)^n (n!)^{-1} \int_C dt_1 \int_C dt_2 \cdots \int_C dt_n \times T_C[H'(t_1)\, H'(t_2) \cdots H'(t_n)]. \tag{5.15}$$

We can apply this expansion directly to the partition function $\mathcal{Q}$ (Eq. 5.6), and likewise to the propagator (Eq. 5.8), where the successive contours connecting τ to t_0, t_0 to t, and t to $\tau - i\beta$ can be joined together to form a single contour which we again call C. The arguments leading to Eq. (4.33) again apply directly, with $\mathcal{Q}/\mathcal{Q}_0$ playing the rôle of $\langle U(\infty, -\infty)\rangle_0$, so that we have, in exact analogy to Eqs. (4.33, 34),

$$G_{AB}(t, t_0) = (\mathcal{Q}/\mathcal{Q}_0)^{-1} \sum_{n=0}^{\infty} (-i)^n (n!)^{-1} \int_C dt_1 \cdots \int_C dt_n \times \langle T_C[H'(t_1) \cdots H'(t_n)\, A(t)\, B(t_0)]\rangle_0; \tag{5.16}$$

$$\mathcal{Q}/\mathcal{Q}_0 = \sum_{n=0}^{\infty} (-i)^n (n!)^{-1} \int_C dt_1 \cdots \int_C dt_n \langle T_C[H'(t_1) \cdots H'(t_n)]\rangle_0. \tag{5.17}$$

The contour C must now (and henceforth) be taken as running from τ to $\tau - i\beta$, and passing through the points t_0 and t, in that order. Alternatively, we may think of C as being fixed, and t_0 and t being chosen as any two points on C with t_0 preceding t along the contour. We can now obtain the generalization of the equation for the causal propagator G^c (Eq. 4.35), by letting t_0 and t be any two points on the fixed contour C, and letting the ordering operator T_C arrange them in the proper order within the product, with a factor of -1 introduced, as usual, if T_C induces an odd permutation of fermion operators.

Our general propagator, then, is

$$G(k, t - t_0) = -i\langle T_C[\mathbf{a}_k(t)\,\mathbf{a}_k^\dagger(t_0)]\rangle \tag{5.18}$$

$$= -i(\mathcal{Q}/\mathcal{Q}_0)^{-1} \sum_{n=0}^{\infty} (-i)^n (n!)^{-1} \int_C dt_1 \cdots \int_C dt_n \times \langle T_C[H'(t_1) \cdots H'(t_n)\, a_k(t)\, a_k^\dagger(t_0)]\rangle_0 . \tag{5.19}$$

Except where t_0 and t pass each other along C, this is an analytic function of $t - t_0$ independent of the choice of contour, provided, of course, the contour passes through both t_0 and t. The domain of analyticity is restricted, as we have seen, by Eq. (5.13), so that the domain of definition is given by

$$-\beta \leqslant \mathrm{Im}\,(t - t_0) \leqslant 0 \tag{5.20}$$

if t_0 precedes t along C, while if t precedes t_0, the domain is

$$0 \leqslant \mathrm{Im}\,(t - t_0) \leqslant \beta . \tag{5.21}$$

If C is to be held fixed while t_0 and t take on all values along C, then these restrictions impose a corresponding restriction on C, namely that a point which moves along C from τ to $\tau - i\beta$ must have a nonincreasing imaginary part. By reasoning almost identical to that leading to Eq. (5.14) we find that this same restriction on C is necessary for the expectation values in Eqs. (5.16, 17) to be defined in general. Possible contours C are shown in Figures 8 (a, b), while Figure 8 (c) shows a contour not allowed.

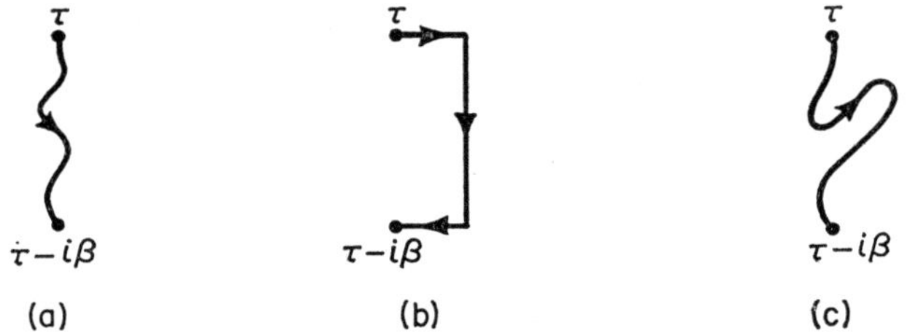

FIGURE 8 Allowed (a, b) and unallowed (c) choices for the contour C, in the complex t plane

The next step in the development of the ground state theory, it will be recalled, was the introduction of Wick's theorem, which, in the interaction representation, permitted the ground state expectation value of a product of annihilation and creation operators to be expressed as a sum of products of simple factors. Perhaps it should not seem surprising that a precisely analogous theorem holds for the grand ensemble statistical expectation values with which we are now confronted; indeed the thermodynamic Wick's theorem has the advantage of being directly applicable to the many-boson system, provided only that the grand ensemble can be used. Like the conventional Wick's theorem, this version is applicable to any of a wide class of ordering operations, including the T_C-ordering of Eqs. (5.17, 19). We use it in the following specific form, analogous to Eq. (4.36) and proved in Appendix B: let $b_1, b_2, \ldots, b_{2p}$ be boson or fermion annihilation or creation operators in the interaction representation, associated with complex times $t_1, t_2, \ldots, t_{2p}$, respectively, along C; and let $\langle(\cdots)\rangle_0$ represent unperturbed statistical expectation values in the grand ensemble. Then

$$\langle T_C(b_1 b_2 \cdots b_{2p})\rangle_0 = \sum_{a.p.p} (-1)^P \prod \langle T_C(b_r b_s)\rangle_0 . \tag{5.22}$$

The sum, as in Eq. (4.36), is over all possible pairings of the factors b_r, and P has the parity of the permutation on fermion operators relating the right to the left side of Eq. (5.22). This has a form identical to Eq. (4.36), and a more precise statement, identical in form to Eq. (4.39), is also possible.

It is now clear that the situation is formally identical to that which we have studied in the ground state problem, Eqs. (5.17, 19, 22) being in just the same form as Eqs. (4.34–36,), and having the same relationships to each other. This means that the analysis in terms of Feynman graphs can be carried out exactly as before, apart from the Fourier transformation in the time variable, which depended on the fact that the domain of t in the ground state theory was the real axis. In the finite temperature theory the new features are that the t integrations run along C from the arbitrary point τ down to $\tau - i\beta$, and the unperturbed propagators, corresponding to the lines of our graphs, are

$$G_0(k, t_i - t_j) = -i \langle T_C[a_k(t_i)\, a_k^\dagger(t_j)] \rangle_0 , \tag{5.23}$$

instead of the propagators $G_0^c(k, t_i - t_j)$ (Eq. 3.37). The explicit form of the analytic propagator $G_0(k, t)$ will not concern us too much, since we shall still find it desirable to work with some sort of Fourier transform. Like G_0, it takes the values $G_0^\pm(k, t_i - t_j)$, depending on the relative values of t_i and t_j, but in this case it is their position along C that matters. If we introduce a function $\theta_C(t_1, t_2)$ defined for t_1 and t_2 on C, and equal to 1 or 0 in case t_1 follows or precedes t_2 along C, respectively, then we can write, in analogy to Eqs. (3.33, 36),

$$\begin{aligned} G_0(k, t_1 - t_2) &= -i\theta_C(t_1, t_2)\, G_0^+(k, t_1 - t_2) \\ &\quad + i\sigma\theta_C(t_2, t_1)\, G_0^-(k, t_1 - t_2) \qquad (5.24) \\ &= -i\, e^{-i\bar{\varepsilon}_k(t_1 - t_2)}[\theta_C(t_1, t_2) f^+(\bar{\varepsilon}_k) - \sigma\theta_C(t_2, t_1) f^-(\bar{\varepsilon}_k)]. \qquad (5.25) \end{aligned}$$

Because of the restriction on C previously discussed, we find that unless $\mathrm{Im}\,(t_1 - t_2) = 0$ we can write

$$\theta_C(t_1, t_2) = \theta\,[-\,\mathrm{Im}\,(t_1 - t_2)], \tag{5.26}$$

and $G_0(k, t_1 - t_2)$ becomes independent of the choice of C.

The prescription for the perturbation expansion can now be taken over bodily; the disconnected graphs give rise now to a factor equal to $\mathcal{Q}/\mathcal{Q}_0$; this cancels the factor $(\mathcal{Q}/\mathcal{Q}_0)^{-1}$ in Eq. (5.19), leaving a prescription for $G(k, t - t_0)$ which may be written in the same symbolic form as Eq. (4.46):

$$G(k, t - t_0) = \sum^{(L)} (-1)^L \int_C \prod dt_i \sum_{\{k_r\}} \prod_i (iv) \prod_r G_0; \tag{5.27}$$

here the factors i and $-i$ have again been rearranged by the use of Eq. (4.49). When a line begins and ends at the same vertex, we must, as before, let the head of the line precede the tail infinitesimally along the path of integration, because it represents the pairing of a factor $a_k^\dagger$ with a factor a_k to the right of it, even though both factors correspond to the time argument. It is sufficient for this purpose to give the corresponding propagator the argument $i\epsilon$, where ϵ is again a positive infinitesimal.

An alternative procedure to that of calculating propagators is to calculate the grand partition function $\mathcal{Q}$ itself which, as we have noted, is given by the sum of all graphs without loose ends (e.g., those in Figure 5). The sum of all such graphs (including a term 1 for no graph at all, the $n = 0$ case) turns out to be the exponential of the sum of those consisting of a single connected part (*not* including the $n = 0$ term). This latter sum then gives the logarithm of the grand partition function, which is proportional to the volume for a large uniform system at fixed density, and is equal to $-\beta\Omega$, where Ω is the thermodynamic potential function (Eq. 3.103).

Before we consider the problem of taking the Fourier transform of the finite-temperature prescription, let us pause to inspect the propagator $G(k, t)$ which we have found it natural to calculate. (We can again take $t_0 = 0$ for convenience, so that C must pass through the origin.) It is defined by Eq. (5.18), and like its unperturbed limit $G_0(k, t)$ (Eqs. 5.23–25), depends ostensibly on the choice of the contour C, but in fact, because of the restrictions on C (Figure 8 and discussion), is independent of C expect on the boundary between the two domains of analyticity, where $\mathrm{Im}\, t = 0$. It is related to the functions $G^{\pm}(k, t)$ (Eqs. 3.12, 13, 56), by

$$G(k, t) = -i\theta\,(-\,\mathrm{Im}\, t)\, G^{+}(k, t) + i\sigma\theta\,(\mathrm{Im}\, t)\, G^{-}(k, t), \tag{5.28}$$

except on the boundary, where $\text{Im}\, t = 0$ and the value given by Eq. (5.28) is ambiguous.

Its spectral representation is obtained from that of $G^{\pm}(k, t)$ (Eq. 3.56), and is

$$G(k, t) = -i \int \varrho(k, \omega)\, e^{-i\omega t} d\omega \,[\theta(-\text{Im}\, t) f^{+}(\bar{\omega}) - \sigma\theta(\text{Im}\, t) f^{-}(\bar{\omega})]\,. \quad (5.29)$$

If $G(k, t)$ is known for all real t, taking the limit from both above and below the real axis, then the spectral function, which is not itself amenable to direct calculation in a perturbation expansion, can be determined by taking the Fourier transform. Indeed the thermodynamic functions $\mathcal{N}$ and E can be determined directly from $G^{-}(k, t)$ in the vicinity of $t = 0$, by the use of Eqs. (3.90, 96), without the necessity for finding the spectral function.

It can be shown, either directly from the definitions (Eqs. 3.12, 13), or from the spectral representation [Eq. (3.56) together with Eq. (3.28)], that $G^{-}(k, t)$ is the same function as $G^{+}(k, t)$ apart from a factor and a simple displacement in t. To be specific,

$$G^{-}(k, t) = e^{\beta\mu} G^{+}(k, t - i\beta)\,. \quad (5.30)$$

As a consequence, $G(k, t)$ itself satisfies a sort of periodicity relation between its two domains of analyticity, which is that if $\text{Im}\, t > 0$,

$$G(k, t) = -\sigma e^{\beta\mu} G(k, t - i\beta)\,. \quad (5.31)$$

5.2 *Analytic self-energy function*

In the perturbation expansion of $G(k, t)$ in terms of Feynman graphs we can identify the self-energy subgraphs, which are connected to one another by a single line, just as in the ground state problem. The sum of these self-energy subgraphs yields a function $M(k, t)$, whose analyticity properties are the same as those of $G(k, t)$; in general however, it includes a contour-dependent delta function at $t = 0$, which may be regarded as the derivative *along* C of a step function discontinuity at the origin. We denote such a contour-dependent delta function by $\delta_C(t - t_0)$ and define it by the properties

$$\delta_C(t - t_0) = 0, \quad t \neq t_0;$$
$$\int_C \delta_C(t - t_0)\, dt = 1\,. \quad (5.32)$$

This delta function is necessary to describe the Hartree–Fock-type graphs shown in Figure 9, in which the initial and final vertices of the graph coincide. These graphs give a contribution of exactly the same form as an external momentum-dependent potential. The rules for calculating $M(k, t - t_0)$ differ

FIGURE 9 The Hartree–Fock-type self-energy graphs, which give a contribution to $M(k, t - t_0)$ proportional to $\delta_C(t - t_0)$. The heavy line represents the true propagator $G(k, i\varepsilon)$ ($i\varepsilon$ is an infinitesimal positive imaginary time), or the entire set of graphs comprising G

from those (Eqs. 5.27) for $G(k, t - t_0)$ only in the absence of initial and final lines and the corresponding factors G_0, and in the additional factor $\delta_C(t - t_0)$ if the initial and final vertices coincide. For future reference, we write this symbolically:

$$M(k, t - t_0) = \sum{}^{(P)} (-1)^L \sum_{\{k_r\}} \int_C \prod_1^{n-2} dt_i \prod_i (iv) \prod_r G_0, \tag{5.33}$$

with $\int_C \prod_1^{n-2} dt_i$ replaced by $\delta_C(t - t_0) \prod_1^{n-1} dt_i$ for Hartree–Fock-type graphs (Figure 9). Here $\sum^{(P)}$ refers to a restriction to "proper" graphs, i.e. those linked graphs which cannot be disconnected by cutting just one particle line.

The self-energy function is related to the propagator by a simple series:

$$\begin{aligned} G(k, t) = G_0(k, t) &+ \int_C dt_1 \, dt_1' \, G_0(k, t - t_1) \, M(k, t_1 - t_1') \\ &\times G_0(k, t_1') + \int_C dt_1 \, dt_1' \, dt_2 \, dt_2' \; G_0(k, t - t_1) \, M(k, t_1 - t_1') \\ &\times G_0(k, t_1' - t_2) \, M(k, t_2 - t_2') \, G_0(k, t_2') + \cdots. \end{aligned} \tag{5.34}$$

This is identical in form, apart from the domain of integration, to what would be obtained by Fourier transforming Eq. (4.69), and, like that series, will be trivial to sum in terms of any sort of Fourier transform of $G(k, t)$ and $M(k, t)$. Thus it will again suffice to find the self-energy function, or some approximation to it.

5.3 Fourier transform—General

A sort of Fourier transform of $G(k, t)$ can be defined by taking the integral

$$\int_C G(k, t)\, e^{izt}\, dt, \tag{5.35}$$

where it should be recalled that C must here pass through the origin, since t_0 has been taken as zero.

If we use the spectral representation (Eq. 5.29) of $G(k, t)$ in this integral, and carry out the t integration explicitly, we see that it takes on a particularly simple form in two special cases, namely if τ is such that

$$e^{iz\tau} = 0, \tag{5.36}$$

or if z is such that

$$e^{\beta(z-\mu)} + \sigma = 0. \tag{5.37}$$

In both of these cases, the integral (5.35) is equal to the analytic propagator $G(k, z)$ defined by Eq. (3.80). The condition (5.36) can be satisfied by letting $\text{Re}\,\tau \to +\infty$ if $\text{Im}\, z > 0$, or by letting $\text{Re}\,\tau \to -\infty$ if $\text{Im}\, z < 0$. Since we can without loss of generality restrict our attention to the upper half of the complex z-plane, we shall refer for the most part only to the first of these, and let C_+ denote a contour C for which $\text{Re}\,\tau \to +\infty$. A typical contour C_+ is indicated in Figure 10(a). In this case we write

$$G(k, z) = \int_{C_+} G(k, t)\, e^{izt}\, dt \qquad (\text{Im}\, z > 0). \tag{5.38}$$

(If the case $\text{Im}\, z < 0$ is to be treated explicitly, one can use a contour C_- [Figure 10(b)], for which $\text{Re}\,\tau \to -\infty$.)

Condition (5.37), on the other hand, is satisfied for the discrete values of z given by

$$z_\nu = \mu + 2\pi i\nu/\beta, \tag{5.39}$$

where for fermions ($\sigma = 1$),

$$\nu = \cdots, -\tfrac{3}{2}, -\tfrac{1}{2}, \tfrac{1}{2}, \tfrac{3}{2}, \cdots, \tag{5.40}$$

while for bosons ($\sigma = -1$),

$$\nu = \cdots, -1, 0, 1, 2, \cdots. \tag{5.41}$$

We can thus write for any of the contours C,

$$G(k, z_\nu) = \int_C G(k, t)\, e^{iz_\nu t}\, dt. \tag{5.42}$$

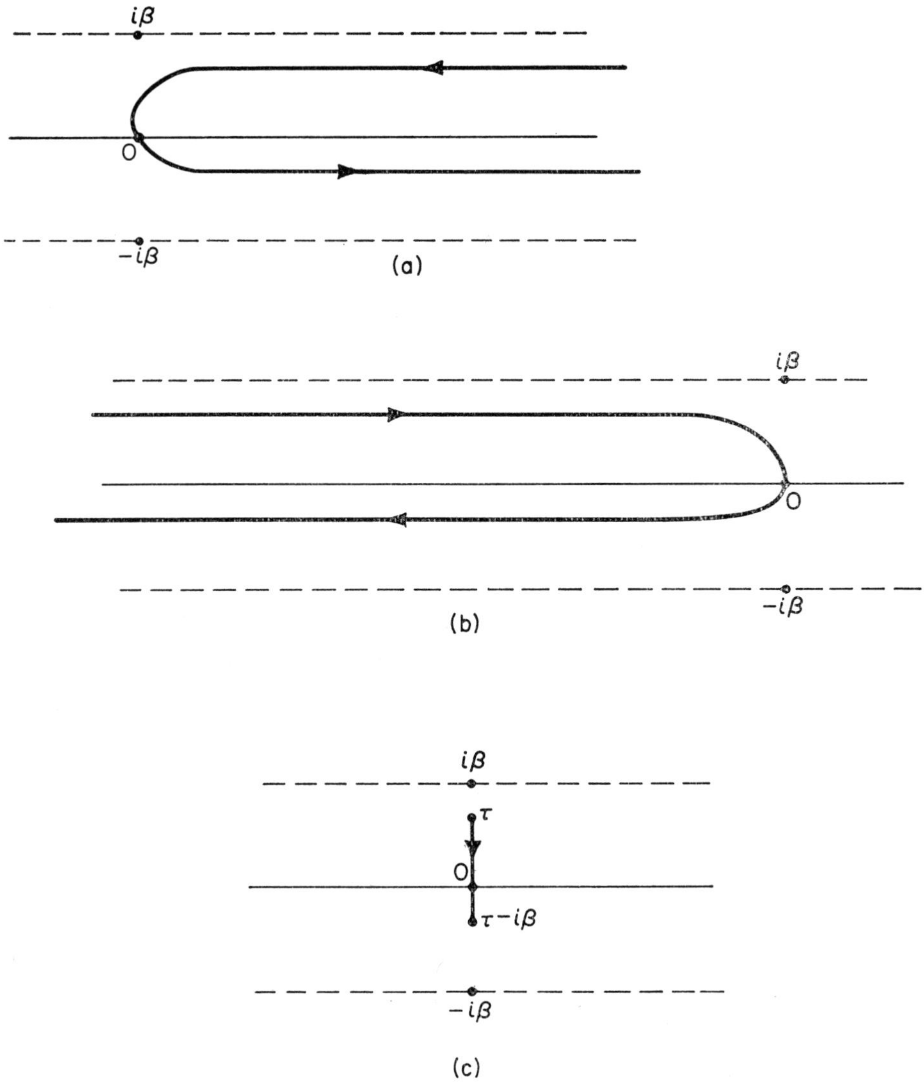

FIGURE 10 The contours (a) C_+, (b) C_-, and (c) C_0, in the complex t plane

These two alternative relations (Eqs. 5.38, 42) form the basis for two of the transformed expansion procedures mentioned earlier. Unfortunately the Feynman graph prescription cannot be taken over directly for the analytic function $G(k, z)$, because the Fourier inversion theorem (Eqs. 3.59, 60) does not hold in general for the modified transform in Eqs. (5.38, 42), and it is the inversion theorem that permits the simple Fourier transformed prescription in the ground state theory. The transformations (5.38, 42) do, however, have some of the properties of the conventional Fourier transform—for example, those relating to the transform of a derivative or of a folding integral. Thus, if $K_i(t)$ are functions of t with the same analyticity and periodicity properties as $G(k, t)$ [e.g., $M(k, t)$], and if one of the restrictions (5.36) or (5.37) holds, then we can define

$$K_i(z) = \int_C K_i(t)\, e^{izt}\, dt. \tag{5.43}$$

For the case $K(t) = M(k, t)$ in particular, this defines an analytic self-energy function $M(k, z)$. If

$$K_2(t) = i\, dK_1(t)/dt, \tag{5.44}$$

say, then we find that

$$K_2(z) = zK_1(z). \tag{5.45}$$

If, alternatively, we have

$$K_3(t) = \int_C K_1(t - t')\, K_2(t')\, dt', \tag{5.46}$$

then the transforms are found to satisfy

$$K_3(z) = K_1(z)\, K_2(z). \tag{5.47}$$

The relation (5.44), incidentally, must be thought of as a differentiation along C, since $K_i(t)$ is in general not analytic along the real axis, and where C touches or crosses the real axis, the derivative is defined only with respect to a path. In particular, if $K_1(t)$ has a discontinuity K_{20}, say, as t passes through zero from above to below the real axis, then $K_2(t)$ has a singularity at the origin of the form $K_{20}\delta_C(t)$ (Eqs. 5.32). From (Eq. 5.43), then, this will contribute a constant K_{20} to $K_2(z)$. In this case the spectral representations of $K_2(t)$ and $K_2(z)$ analogous to (5.29) and (3.80) must display the constant term explicitly, and thus have the forms

$$K_2(t) = K_{20}\delta_C(t) - i \int \varkappa_2(\omega)\, e^{-i\omega t} d\omega\, [\theta_C(t, 0) f_2^+(\omega) - \sigma_2 \theta_C(0, t) f_2^-(\omega)], \tag{5.48}$$

$$K_2(z) = K_{20} + \int_{-\infty}^{\infty} \frac{\varkappa_2(\omega)}{z - \omega}\, d\omega, \tag{5.49}$$

where the spectral function $\varkappa_2(\omega)$ can be obtained from $K_2(z)$ by

$$\varkappa_2(\omega) = -(2\pi i)^{-1} [K_2(\omega + i\eta) - K_2(\omega - i\eta)]. \tag{5.50}$$

A third and less obvious relationship* is obtained for three functions related by

$$K_3(t) = K_1(t)\, K_2(t). \tag{5.51}$$

In this case, K_1 and K_2 may describe different types of particle or particle combinations, and K_3 in turn describes the combination of those associated with K_1 and K_2. Thus there will in general be three different chemical potentials, related by

$$\mu_3 = \mu_1 + \mu_2, \tag{5.52}$$

and three different statistical parameters σ (Eq. 3.22), related by

$$\sigma_3 = -\sigma_1\sigma_2. \tag{5.53}$$

The functions K_1 and K_2 may not be too singular; for example, if one of them has a term proportional to $\delta_C(t)$, then the other must be continuous at $t = 0$. If $\varkappa_1(\omega)$ and $\varkappa_2(\omega)$ are the spectral functions (Eq. 5.48–50) for K_1 and K_2, respectively, then $K_3(z)$ is given (among a variety of equivalent forms) by

$$K_3(z) = -\tfrac{1}{2}i \int_{-\infty}^{\infty} d\omega_1 \alpha_1(\bar{\omega}_1)\, \varkappa_1(\omega_1)\, K_2(z - \omega_1)$$
$$-\tfrac{1}{2}i \int_{-\infty}^{\infty} d\omega_2\, \alpha_2(\bar{\omega}_2)\, \varkappa_2(\omega_2)\, K_1(z - \omega_2), \tag{5.54}$$

where $\alpha_i(x)$ is determined by Eq. (3.71), with the appropriate values σ_i and μ_i. Another useful form for this relationship is the expansion for the spectral function $\varkappa_3(\omega)$:

$$\varkappa_3(\omega_3) = -\tfrac{1}{2}i \int d\omega_1 \left[\alpha_1(\bar{\omega}_1) + \alpha_2(\bar{\omega}_3 - \bar{\omega}_1)\right] \varkappa_1(\omega_1)\, \varkappa_2(\omega_3 - \omega_1). \tag{5.55}$$

This does not determine the value of a possible constant term K_{30} [Eqs. (5.48, 49)], which is easily determined, however, as the coefficient of $\delta_C(t)$ in Eq. (5.51). These relationships (5.54, 55) can be readily obtained by substituting the spectral forms (5.48) for K_1 and K_2 (K_{10} and K_{20} cannot both differ from zero, though) into (5.51) and (5.43) and performing the integrations. The identity

$$f_1^+(\bar{\omega}_1) f_2^+(\bar{\omega}_2) - \sigma_1\sigma_2 f_1^-(\bar{\omega}_1) f_2^-(\bar{\omega}_2) = \tfrac{1}{2}\left[(\alpha_1(\bar{\omega}_1) + \alpha_2(\bar{\omega}_2)\right], \tag{5.56}$$

which is easily checked, is useful in getting the results into the forms (5.54, 55).

The relations (5.46, 47) permit the summing of the series in Eq. (5.34), which expresses $G(k, t)$ in terms of $M(k, t)$. This becomes, in fact, a simple

---

* See note on page 106.

geometric series, similar to Eq. (4.69), involving $M(k, z)$ and $G_0(k, z)$, and easily summed. Using the fact (Eqs. 3.80, 87) that

$$G_0(k, z) = (z - \varepsilon_k)^{-1}, \tag{5.57}$$

we obtain simply

$$G(k, z) = [z - \varepsilon_k - M(k, z)]^{-1}, \tag{5.58}$$

in close analogy with Eq. (4.70). This holds also for $\operatorname{Im} z < 0$, and $M(k, z)$ is seen, like $G(k, z)$, to have a cut at the real axis. Since $G(k, z)$ has no zeros off the real axis [this follows from Eq. (3.80), whose real and imaginary parts cannot both be zero off the real axis], $M(k, z)$ has no poles, and is thus analytic in the cut plane.

Just as in the ground state problem, we can see that it is generally more fruitful to calculate the self-energy function in perturbation theory than the propagator itself. Note, in particular, that in any finite order in the perturbation expansion of $G(k, z)$, the factor $G_0(k, z)$ will appear to some power—at least squared, except in zero order—so that such an approximation to $G(k, z)$ has a higher-order pole in z at the unperturbed energy ε_k, while the true $G(k, z)$ has in general no singularity at all at the unperturbed energy ε_k.

It is tempting to look for poles in $G(k, z)$ by finding the roots of the denominator, which would be expected to be complex since in general $M(k, z)$ is complex for real z. However, $G(k, z)$ has no poles in the cut z-plane, and there is no guarantee that the function $M(k, z)$ has an analytic continuation across the real axis to unphysical sheets, or if it does, that the function $z - \varepsilon_k - M(k, z)$ has zeros rather than cuts or other types of singularity.

5.4 *Imaginary-time procedure*

We return now to the problem of the Feynman graph prescription for $G(k,z)$ or $M(k, z)$. The first procedure we consider, which has been used by a number of authors, is to take τ pure imaginary, and C lying entirely on the imaginary t-axis. We call this contour C_0 (see Figure 10c). Then $G(k, t)$ satisfies a periodicity relation on C_0 (Eq. 5.31), which permits one to express it as a conventional Fourier series, and to identify $G(k, z_\nu)$ (Eq. 5.42) as the Fourier coefficient. [It is in fact $G(k, t)\, e^{i\mu t}$ which is periodic for Bose statistics and antiperiodic for Fermi statistics; this accounts for the constant μ in the definition of z_ν.] The Fourier expansion of $G(k, t)$, for pure imaginary t, is then

$$G(k, t) = i\beta^{-1} \sum_\nu G(k, z_\nu)\, e^{-iz_\nu t}, \tag{5.59}$$

with z_ν given by Eqs. (5.39–41). Exactly similar relations hold for $G_0(k, t)$ and for the self-energy function $M(k, t)$. We are now in an analogous position to that in the ground state problem which enabled us to state the Feynman graph expansion in terms of energy integrals rather than time integrals. The same analysis carries through, therefore, with just one distinctive difference, apart from the fact that the factors are different, namely that sums over ν replace the integrals over ω. The integral along C_0, with respect to a t variable, associated with each vertex gives rise to a Kronecker delta in the indices ν analogous to the energy conservation delta function of the ground state theory. This means that the sum of the ν's associated with lines entering any vertex is equal to the sum of the ν's leaving the vertex. The same is true of the corresponding values of z_ν, because of the linear relation (Eq. 5.39) between z_ν and ν; the term μ in this relation causes no trouble because of particle number conservation—that is, as many lines leave as enter any vertex, so that the terms μ just balance each other. (Again, if particle number is not conserved, $\mu = 0$.)

The net result is the following prescription for calculating $G(k, z_\nu)$: the graphs are drawn and enumerated according to the same rules as before, with n vertices for an nth-order graph, in addition to the two endpoints; each line is labeled with a momentum k_ν and an index ν_r (or the complex energy z_{ν_r}), and corresponding to each line is a factor $G_0(k_\nu, z_{\nu_r})$ and a sum over k_r and ν_r (Eqs. 5.40, 41); corresponding to each vertex are factors $-1/\beta$, the appropriate matrix element of v, and a Kronecker delta of the form $\delta(\nu_r + \nu_s, \nu_{r'} + \nu_{s'})$. The factor $-1/\beta$ for each vertex is obtained by collecting factors and using the relation (4.49) in the same way as was done previously with factors of 2π and i; indeed one finds quite generally that the factor 2π associated with the Fourier integral is here replaced everywhere by a factor $-i\beta$ wherever it occurs. A line that begins and ends at the same vertex gives rise to a factor $e^{\varepsilon z_\nu}$, arising, through Eq. (5.59), from the time argument $i\varepsilon$ for such lines, previously mentioned. This factor provides convergence for the corresponding sum over ν, which is otherwise logarithmically divergent. There is again a factor -1 for each closed fermion loop. The symbolic form for this prescription, analogous to Eq. (4.50), is

$$G(k, z_\nu) = \sum^{(L)} (-1)^L \sum_{\{k_r\}} \sum_{\{\nu_r\}} \prod_i [-\beta^{-1} v(r, s; r', s')\, \delta(\nu_r + \nu_s, \nu_{r'} + \nu_{s'})]$$
$$\times \prod_r [G_0(k_r, z_{\nu_r}) \exp(\varepsilon z \nu_r)]. \qquad (5.60)$$

The recipe for calculating the self-energy function $M(k, z_\nu)$ is the same [see Eq. (S. 33)] except for the restriction to proper graphs and the omission of the two and lines, together with the corresponding factor $[G_0(k, z_\nu)]^2$.

The thermodynamic functions $\mathcal{N}$ and E can readily be computed once the propagator is known for the discrete values z_ν of the arguments; from Eqs. (3.90, 96) they are given in terms of the function $G^-(k, t)$ at $t = 0$, which in turn can be replaced by $-i\sigma G(k, t)$ (Eq. 5.28) if we let $t = i\varepsilon$. Using, finally, Eq. (5.59) to express things in terms of $G(k, z_\nu)$, we get

$$\mathcal{N} = \sigma\beta^{-1} \sum_{k,\nu} G(k, z_\nu)\, e^{\varepsilon z_\nu}; \tag{5.61}$$

$$E = \tfrac{1}{2}\sigma\beta^{-1} \sum_{k,\nu} (z_\nu + \varepsilon_k)\, G(k, z_\nu)\, e^{\varepsilon z_\nu}. \tag{5.62}$$

The summand is not real in these two expressions, but it is readily seen (Eqs. 3.80, 5.39) that the $(-\nu)^{\text{th}}$ term in each sum is the complex conjugate of the νth term, so that the sum is in fact real.

It is not so straightforward to find the spectral function from the propagator $G(k, z_\nu)$, although it is in fact determined, together with the whole analytic function $G(k, z)$, once these discrete values are known [57]. In the first place $G(k, z_\nu)$ determines $G(k, t)$ for pure imaginary t by Eq. (5.59); from this the entire analytic function $G(k, t)$ is determined by analytic continuation, and hence, by transformation back (Eq. 5.38), the analytic function $G(k, z)$ is determined, and the spectral function can, of course, be obtained by taking the imaginary part of $G(k, z)$ for real z. Alternatively, $G^{\pm}(k, t)$ can be determined from $G(k, t)$ (Eq. 5.28), and these in turn used, with t real (Eqs. 3.60, 61), to obtain $\varrho(k, \omega)$. Even if this procedure could be carried out, however, any slight error in the values $G(k, z_\nu)$ would destroy the analyticity properties of $G(k, t)$, so that as a numerical procedure this is not recommended. For many approximations of practical interest it will be found possible to perform these steps, or their equivalent, analytically, and so perform direct calculations for real values of the energy z.

5.5 *Analytic transform*

The second approach to a Fourier transformed theory which we present here is due to Dzyaloshinskii [55], and Baym and Sessler [56], though our derivation, which makes use of the relation (5.38) involving the contour C_+, differs somewhat from that of Baym and Sessler, which uses the contour C_0. We consider the self-energy function $M(k, z)$ rather than the propagator because of certain complications with vanishing energy denominators which arise in the case of $G(k, z)$, and which we shall discuss briefly later. We use the transformation for $M(k, z)$ of the same kind as Eq. (5.38),

$$M(k, z) = \int_{C_+} M(k, t)\, e^{izt}\, dt \quad (\operatorname{Im} z > 0), \tag{5.63}$$

and substitute on the right the expansion (Eq. 5.33) for $M(k, t)$, with t_0 again set equal to zero, and the contour C_+ used for all t integrations. The contribution of a given n^{th}-order graph to $M(k, z)$ is thus expressed as an in-

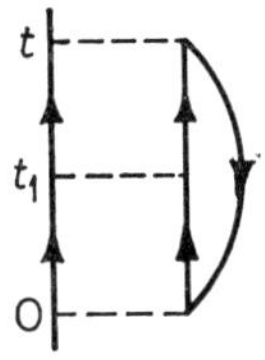

FIGURE 11 Typical third-order self-energy graph

tegral over the $n - 1$ variables $t_1, t_2, \ldots, t_{n-2}$, and t, and can be regarded as the sum of contributions corresponding to the different orderings of these variables along C_+ with respect to each other and with respect to the origin. These different orderings can be indicated graphically by using time-ordered graphs, in which the vertices are ordered vertically, say, in the same way that the corresponding variables are ordered along C_+. Thus the single self-energy graph shown in Figure 11, for example, would represent the sum of the six time-ordered graphs of Figure 12. That shown in Figure 12(b), for

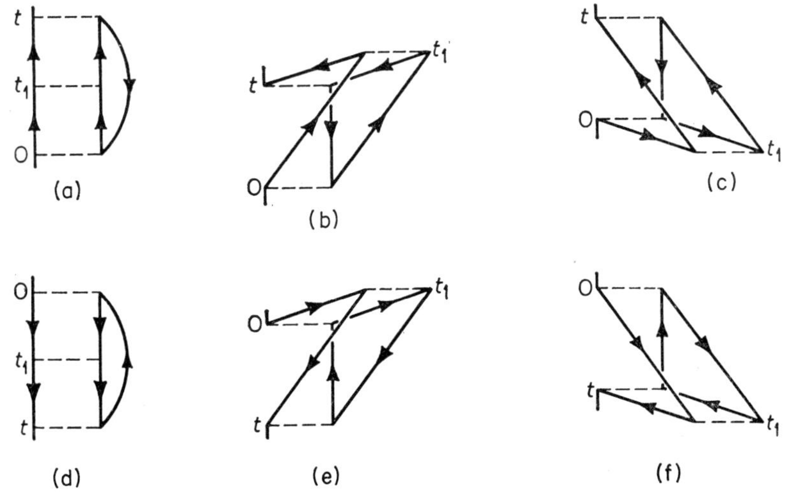

FIGURE 12 Time-ordered self-energy graphs corresponding to Figure 11

instance, would correspond [through Eqs. (5.33, 63)] to the following contribution to $M(k, z)$ for a fermion system:

$$M_{12(b)}(k, z) = -\int_{\substack{0 \\ C_+}}^{\tau - i\beta} dt \int_{\substack{t \\ C_+}}^{\tau - i\beta} dt_1 e^{izt} \sum_{k_1 k_2 K} [iv(k, K-k; k_1, K-k_1)]$$
$$\times [iv(k_1, K-k_1; k_2, K-k_2)] [iv(k_2, K-k_2; k, K-k)]$$
$$\times G_0(k_1, t-t_1) G_0(k_2, t_1) G_0(K-k_1, t-t_1)$$
$$\times G_0(K-k_2, t_1) G_0(K-k, -t) \quad (\mathrm{Re}\,\tau \to +\infty), \qquad (5.64)$$

which, by virtue of the time ordering and Eq. (5.25), reduces to

$$M_{12(b)}(k, z) = -\sum_{k_1 k_2 K} v(k, K-k; k_1, K-k_1)$$
$$\times v(k_1, K-k_1; k_2, K-k_2)\, v(k_2, K-k_2 : k, K-k) f^-(\bar{\varepsilon}_{k_1})$$
$$\times f^+(\bar{\varepsilon}_{k_2}) f^-(\bar{\varepsilon}_{K-k_1}) f^+(\bar{\varepsilon}_{K-k_2}) f^-(\bar{\varepsilon}_{K-k}) \int_{\substack{0 \\ C_+}}^{\tau - i\beta} dt \int_{\substack{t \\ C_+}}^{\tau - i\beta} dt_1\, e^{izt}$$
$$\times \exp[-i\varepsilon_{k_1}(t-t_1) - i\varepsilon_{k_2}t_1 - i\varepsilon_{K-k_1}(t-t_1) - i\varepsilon_{K-k_2}t_1 + i\varepsilon_{K-k}t]$$
$$(\mathrm{Re}\,\tau \to +\infty). \qquad (5.65)$$

In a similar way, the contribution corresponding to Figure 12(c) turns out to be

$$M_{12(c)}(k, z) = -\sum_{k_1 k_2 K} v(k, K-k; k_1, K-k_1)$$
$$\times v(k_1, K-k_1; k_2, K-k_2)\, v(k_2, K-k_2; k, K-k) f^+(\bar{\varepsilon}_{k_1}) f^-(\bar{\varepsilon}_{k_2})$$
$$\times f^+(\bar{\varepsilon}_{K-k_1}) f^-(\bar{\varepsilon}_{K-k_2}) f^-(\bar{\varepsilon}_{K-k}) \int_{\substack{0 \\ C_+}}^{\tau - i\beta} dt \int_{\substack{\tau \\ C_+}}^{0} dt_1\, e^{izt}$$
$$\times \exp[-i\varepsilon_{k_1}(t-t_1) - i\varepsilon_{k_2}t_1 - i\varepsilon_{K-k_1}(t-t_1) - i\varepsilon_{K-k_2}t_1 + i\varepsilon_{K-k}t]$$
$$(\mathrm{Re}\,\tau \to +\infty). \qquad (5.66)$$

It will be seen that the expressions (5.65) and (5.66) are very similar in detail, differing only in the ranges of integration and in the factors $f^{\pm}(\bar{\varepsilon})$. The same is true for all of the graphs of Figure 12, except that the overall sign also may vary. Our prescription for $M(k, z)$ is obtained by simply doing the t integrations explicitly and arranging the factors in a convenient way. Those integrations which are not already convergent on account of the complex z can be made to converge to the proper generalized function by the use of a factor $e^{\pm\eta t i}$ where η is a positive infinitesimal, as usual. The double integral in Eq. (5.65), for instance, gives a factor

$$[i(\varepsilon_{k_1} + \varepsilon_{K-k_1} - \varepsilon_{k_2} - \varepsilon_{K-k_2} + i\eta)^{-1}]\,[i(z + \varepsilon_{K-k} - \varepsilon_{k_2} - \varepsilon_{K-k_2})^{-1}], \quad (5.67)$$

while the integral in Eq. (5.66) gives

$$[i(z + \varepsilon_{K-k} - \varepsilon_{k_1} - \varepsilon_{K-k_1})^{-1}]\,[i(\varepsilon_{k_2} + \varepsilon_{K-k_2} - \varepsilon_{k_1} - \varepsilon_{K-k_1} - i\eta)^{-1}]. \quad (5.68)$$

These denominators can be related to the energies obtained by cutting the graph horizontally between successive pairs of vertices and adding the free-particle energies of the lines so cut, with a plus sign for downward-directed lines and a minus sign for upward-directed lines. The appearance of z in these denominators can be handled by the same rule if an additional line, carrying energy z, is imagined drawn from the vertex at time t back to the vertex at time zero. This is illustrated for the graph of Figure 12(b), in Figure 13.

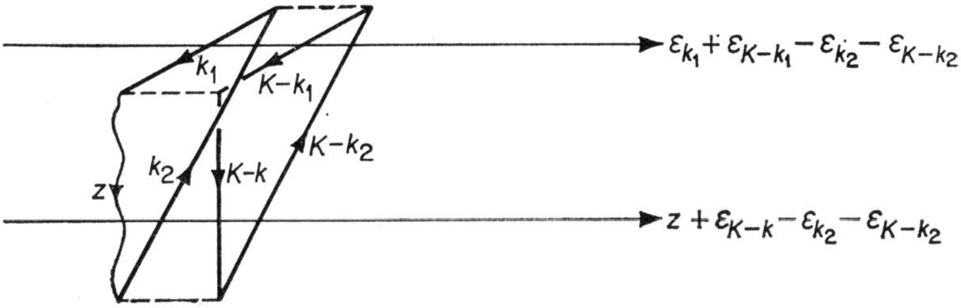

FIGURE 13 Illustrating the calculation of energy denominators for the self-energy graph 12(b)

This procedure for calculating the energy denominators is found to be valid for the most general time-ordered self-energy graph, by simply doing the various t integrations in turn, starting with those variables nearest the ends of C_+ (at the top and bottom of the graph) and working in to those nearest the fixed vertex at the time zero. When the various factors are collected, we have the following rules: all possible self-energy graphs are drawn, with all possible orderings of vertices from top to bottom of each graph. Each line is labeled with a momentum and the corresponding free-particle energy; an additional line is drawn between the end points of the graph, bearing the momentum k and the complex energy z. Momentum is conserved at each vertex. Horizontal cuts are made between successive vertices from top to bottom of the graph, and for each cut α the corresponding energy denominator D_α is calculated by adding the energies of downward-directed lines cut and subtracting those of upward-directed lines cut. A term $i\eta$ is added for cuts above the upper endpoint and subtracted for cuts below the lower endpoint. (For cuts between the endpoints, the denominator is complex and so there is no problem.) For each vertex there is a factor equal to the appropriate matrix element of v. For each upward-directed line there is a factor $f^+(\bar{\varepsilon}_{k_r})$, and for each downward-directed line, a factor $-\sigma f^-(\bar{\varepsilon}_{k_r})$. For each cut there is a factor D_α^{-1}, as calculated above, and for each closed fermion loop a factor -1. A line which begins and ends at the same vertex is regarded as downward-directed. All internal momenta are summed over. Symbolically, we write

$$M(k, z) = \sum\nolimits^{(P)} (-1)^L \sum_{\{k_r\}} \prod_i v \prod_\alpha D_\alpha^{-1} \prod_r{}^{(1)} f^+ \prod_r{}^{(2)} (-\sigma f^-), \tag{5.69}$$

where $\prod_r^{(1)}$ signifies a product over upward directed lines, and $\prod_r^{(2)}$, over downward directed lines.

The same analysis can be carried through with respect to the argument t_0, the argument t being set equal to zero; if we want to keep $\operatorname{Im} z > 0$, we must use the contour C_-:

$$M(k, z) = \int_{C_-} M(k, -t_0)\, e^{-izt_0} dt_0 \quad (\operatorname{Im} z > 0). \tag{5.70}$$

The result of this analysis is a prescription identical to Eq. (5.69), but with $i\eta$ replaced by $-i\eta$ in the energy denominators. One finds, in other words, that only the relative signs of the imaginary infinitesimals are of importance in the prescription. The prescription for $\operatorname{Im} z < 0$ is obtained by using the contour C_- in Eq. (5.63) or C_+ in Eq. (5.70). The effect of this is again simply to change the sign of the imaginary infinitesimal, which we have already seen to be immaterial anyway. Thus the prescription (5.69) is valid for arbitrary complex z, and with arbitrary sign for η.

A certain difficulty arises if the graph being calculated contains a self-energy part as a subgraph, because then it is possible for one of the horizontal cuts to cut just two lines, one up and one down, with the same energy, and so to give a zero denominator. An example of such an "anomalous graph" is shown in Figure 14, where the cut shown gives a zero denominator. Such anomalous graphs could not have been avoided if we had calculated $G(k, z)$ by this technique, which is why we did not try to do so. We can avoid the problem in the calculation of $M(k, z)$ by using $G(k, t)$ instead of $G_0(k, t)$ for the internal lines, and considering only irreducible graphs, that is, those with no self-energy subgraphs. These propagators $G(k, t)$, for which we use the spectral representation (5.29), can be calculated in some simpler approximation, or can be regarded as determined self-consistently in terms of the self-energy function $M(k, z)$ which we are calculating. The spectral representation (5.29) shows the same form of dependence on its time argument as

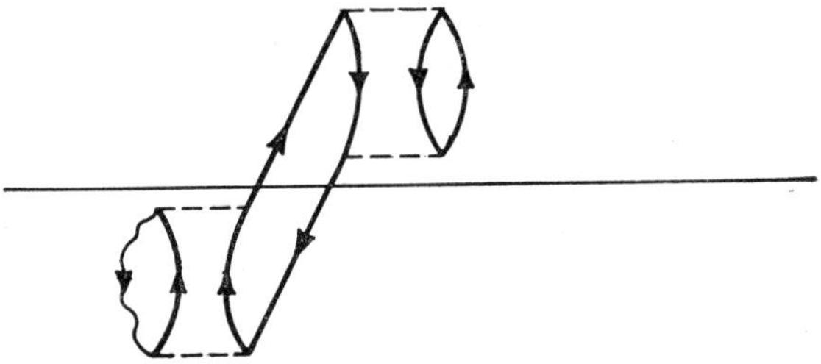

FIGURE 14 An anomalous graph. The cut indicated corresponds to a zero energy denominator

the unperturbed propagator $G_0(k, t)$ (Eq. 5.25) except that the free-particle energy is replaced by an integration variable ω, and the spectral function $\varrho(k, \omega)$ is introduced into the integral. The reasoning leading to the prescription (5.69) remains unaltered, and the resulting rules are modified only in these points: only irreducible graphs are counted; each line is assigned an energy variable ω_r, which is to be integrated over, independent of its momentum k_r, and a factor $\varrho(k_r, \omega_r)$, which is the particle spectral function or some approximation to it; the factors $f^{\pm}(\bar{\varepsilon}_k)$ are replaced by $f^{\pm}(\bar{\omega}_r)$, and the energy denominators, calculated in the same way, now involve the variables ω_r instead of the free-particle energies ε_{k_r}. Symbolically, the modified prescription is

$$M(k, z) = \sum^{(I)}(-1)^L \sum_{\{k_r\}} \int \prod d\omega_r \prod_r \varrho(k_r, \omega_r)$$

$$\times \prod_i v \prod_\alpha D_\alpha^{-1} \prod_r^{(1)} f^+(\bar{\omega}r) \prod_r^{(2)} [-\sigma f^-(\bar{\omega}_r)]. \quad (5.71)$$

The approach we have been describing, which resembles the Goldstone expansion for the ground state energy, differs from the first expansion we discussed in that the lines of each graph do not correspond to unperturbed propagators as factors in the calculation; rather, the graph represents a succession of intermediate states, connected by appropriate matrix elements as in stationary-state perturbation theory, and associated in like manner with energy denominators in the expression to be calculated. Two advantages of this approach are that the rules are comparatively simple and that one is calculating directly the analytic self-energy function $M(k, z)$. This latter point means that one may set z real (with positive or negative infinitesimal imaginary part) and so find the spectral function directly without further analytic continuation. One disadvantage is that the different time-orderings of the vertices (along C) must be enumerated explicitly, which means that a much larger number of terms must be calculated in any order, and that one cannot easily use summation procedures which treat particle lines and hole lines on an equal basis (e.g., summing ladder graphs of the type shown in Figure 11, with an arbitrary number of interactions and including all time orderings of the vertices). Such a sum, feasible by the other approaches we consider, does not seem practicable by this technique.

5.6 *Real-time procedure*

The third approach to a Fourier transformed prescription which we consider is to start with a real-time theory [58], obtained by distorting the contour C to include the entire real-time axis. The Fourier transform can then be taken with respect to real values of t, so that we end up with an expansion involving only real energy variables. The contour may be chosen in several ways, of which the one we shall use, which we call C_R, is shown in Figure 15.

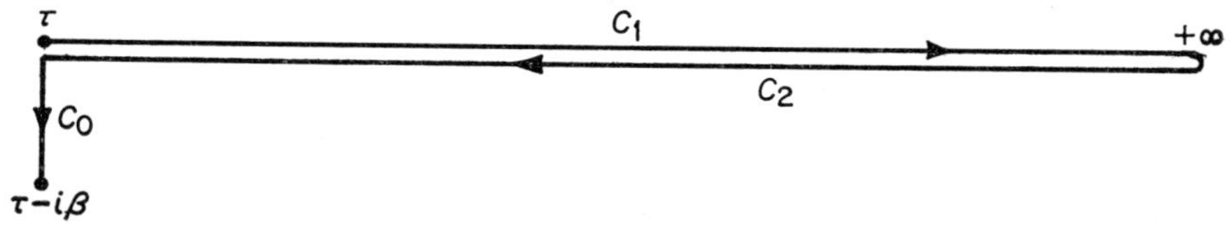

FIGURE 15 The contour C_R in the complex t plane; $\tau \to -\infty$, and C_1 and C_2 lie on the real axis

The initial point τ is allowed to approach $-\infty$ ($\text{Im}\,\tau = 0$), and the contour runs along the real axis to $t = +\infty$, back to $t = \tau$, and then down to $\tau - i\beta$. These three portions are labeled C_1, C_2, and C_0, as indicated in the figure. We can use this contour in the recipe of Eq. (5.27), or the corresponding recipe (5.33) for the self-energy function $M(k, t - t_0)$, provided the two times t and t_0 lie on C_R. In fact, to keep them finite we must restrict our attention to the case that t and t_0 both lie on the two legs C_1 and C_2. It is the presence of these fixed arguments that make this choice of contour nontrivially different from that consisting of C_0 alone, with τ finite; if t and t_0 both lay on C_0, or if we were calculating the partition function, which has no fixed time arguments, then the contributions of C_1 and C_2 would exactly cancel.

As it is, however, the points on C_0 are taken infinitely far away from the fixed points t and t_0 in the limit, so that if the t integrations converge at infinite t, the contribution of this finite segment is negligible and we have a real-time theory. Actually the time dependence of the integral comes from the factors $G_0(k, t_i - t_j)$ in Eqs. (5.27, 33), and is simply oscillatory (Eq. 5.25); in most cases the momentum integration smooths this out (in the infinite volume limit—at finite volume the problem is more complicated) and the contribution of C_0 does indeed vanish. Some trouble arises for a graph containing one or more self-energy subgraphs (connected to the rest of the graph by just two lines carrying the same momentum). Then it is possible, if the vertices of the subgraph all lie on C_0, and no other vertices do, for the integral not to be oscillatory in τ at all as τ becomes infinite, so that no smoothing out can take place. This trouble is closely akin to the anomalous graph problem encountered in the direct calculation of $M(k, z)$, and can be dealt with in the same way, namely, by calculating only the irreducible self-energy graphs, with G_0 replaced by G (or some approximation to it) for every line of each graph.

A more careful analysis shows that in fact the contribution of C_0 can be ignored throughout the derivation, provided one ends up by using the irreducible self-energy graphs, with G instead of G_0 for internal lines. (It is all right to use G_0 as an approximation to G, but the graphs must be irreducible.) The actual perturbation series can then be deduced from this, if desired, by making a simple power series expansion of each factor in each term. As pointed out in the previous section, however, the terms of the actual perturbation series for the propagator or spectral function are highly singular at the free-particle energy, while the true functions are in general not singular there at all. This fact is at the root of the anomalous graph difficulty, and explains why it is desirable to use the irreducible self-energy graph procedure.

We therefore proceed by supposing that C_0 can be ignored throughout, so that every variable t that arises is on C_1 or C_2. Our variables are therefore real, and we can use an index $a = 1, 2$, to indicate which leg of the contour, C_1 or C_2, each point lies on. We are thus led to define a 2×2 matrix propagator $\Gamma(k, t - t_0)$ by

$$\Gamma_{ab}(k, t - t_0) = G(k, t - t_0)\, \sigma_b, \tag{5.72}$$

where $G(k, t - t_0)$ is defined with respect to the contour C_R, and where

$$t \in C_a; \quad t_0 \in C_b; \tag{5.73}$$

$$\sigma_1 = 1; \quad \sigma_2 = -1. \tag{5.74}$$

The minus sign for $b = 2$ is introduced to simplify the form of various relations satisfied by Γ, and by the corresponding self-energy matrix Σ to be defined later; it is associated with the necessity of changing the direction of integration over C_2. The four components of Γ are the familiar functions G^c, $G^{\pm}$, and $G^{\sim c}$ (Eqs. 3.12, 13, 34, 38), as follows simply from the definitions (Eqs. 5.72, 18) and the arrangement of the contour C_R:

$$\Gamma(k, t) = \begin{bmatrix} G^c(k, t) & -i\sigma G^-(k, t) \\ -iG^+(k, t) & G^{\sim c}(k, t) \end{bmatrix} \tag{5.75}$$

Analogous matrix (or tensor) functions can be defined for a wide class of functions which can be defined along our arbitrary C, by simply specializing to the contour C_R, and using the index a to indicate the legs C_1 and C_2. In particular, the self-energy function $M(k, t - t_0)$ specializes to a self-energy matrix function Σ:

$$\Sigma_{ab}(k, t - t_0) = M(k, t - t_0)\,\sigma_b, \tag{5.76}$$

where again we use Eqs. (5.73, 74).

Now we already have a prescription (Eq. 5.27, 33) which can be used to calculate $G(k, t - t_0)$ or $M(k, t - t_0)$ in perturbation theory, using any of the contours C, or in particular, using C_R. Each integration along C_R reduces to an integration along the real t axis and a summation over the two values of the index a, with a minus sign for $a = 2$, provided by a factor σ_a at each vertex, to compensate for the change in direction of integration along C_2. The resulting prescription is very similar to that for $G(k, t - t_0)$: to each vertex i we assign an index a_i, in addition to the time variable t_i; the symbolic prescription is

$$\Gamma_{ab}(k, t - t_0) = \Sigma^{(L)}(-1)^L \sum_{\{a_i\}} \sum_{\{k_r\}} \int_{-\infty}^{\infty} \prod dt_i \times \prod_i (i\sigma_{a_i} v) \prod_r \Gamma_0(k_r, t_i - t_j). \tag{5.77}$$

For a line that begins and ends at the same vertex, we have found previously that the head of the line must be regarded as lying infinitesimally farther back along C than the tail. This means here that the time argument of the corresponding factor Γ_{0aa} must be $-\epsilon$ for $a = 1$, and $+\epsilon$ for $a = 2$, ϵ being a positive infinitesimal as before. We can therefore let the argument of the propagator be $-\sigma_a \varepsilon$ in this case.

The prescription for the self-energy matrix $\Sigma(k, t - t_0)$ is [like that for $M(k, t - t_0)$] exactly the same as that for the propagator, except for an additional factor, in this case $\delta_{ab}\,\delta(t - t_0)$, in the case of the Hartree–Fock-type graphs (Figure 9). The propagator is related to the self-energy matrix by a relation of the same form as Eq. (5.34):

$$\Gamma(k, t - t_0) = \Gamma_0(k, t - t_0) + \int dt_1 dt_1'\, \Gamma_0(k, t - t_1) \times \\ \times \Sigma(k, t_1 - t_1')\, \Gamma_0(k, t_1' - t_0) + \cdots, \quad (5.78)$$

where matrix multiplication is understood, and the factors σ_b in the definitions (Eq. 5.72, 76) of Γ and Σ have been used to compensate for the change in direction of integration along C_2.

We now have a complete real-time theory; it remains only to take the Fourier transform in the same manner as for the ground state theory, where we went from the prescription (4.46) for $G^c(k, t)$ to the prescription (4.50) for $G^c(k, \omega)$. Following the general definition (Eq. 3.60) for the Fourier transform, we take

$$\Gamma_{ab}(k, \omega) = \int_{-\infty}^{\infty} \Gamma_{ab}(k, t)\, e^{i\omega t}\, dt, \quad (5.79)$$

where ω must now be real. From the expression (Eq. 5.75) for Γ in terms of the simpler propagators, and from their expression (Eqs. 3.74, 77, 79) in terms of $G^R(k, \omega)$ and $G^A(k, \omega)$, we find that

$$\Gamma(k, \omega) = \begin{bmatrix} G^c(k, \omega), & -i\sigma G^-(k, \omega) \\ -iG^+(k\,\omega) & G^{\sim c}(k, \omega) \end{bmatrix} \quad (5.80)$$

$$= \begin{bmatrix} f^+(\bar\omega) & \sigma f^-(\bar\omega) \\ f^+(\bar\omega) & \sigma f^-(\bar\omega) \end{bmatrix} G^R(k, \omega)$$

$$+ \sigma \begin{bmatrix} f^-(\bar\omega) & -f^-(\bar\omega) \\ -\sigma f^+(\bar\omega) & \sigma f^+(\bar\omega) \end{bmatrix} G^A(k, \omega)$$

$$= \varphi^+(\bar\omega)\, G^R(k, \omega) + \sigma\varphi^-(\bar\omega)\, G^A(k, \omega), \quad (5.81)$$

where we have let

$$\varphi^+ = \begin{bmatrix} f^+ & \sigma f^- \\ f^+ & \sigma f^- \end{bmatrix}; \\ \varphi^- = \begin{bmatrix} f^- & -f^- \\ -\sigma f^+ & \sigma f^+ \end{bmatrix}. \quad (5.82)$$

This form (Eq. 5.81) is in striking analogy to the form (3.74) for $G^c(k, \omega)$, especially when the latter is considered in the $T = 0$ limit, where $G^c(k, \omega)$ plays the rôle that $\Gamma(k, \omega)$ does in the finite-temperature theory. This is because the matrices φ^+ and $\sigma\varphi^-$ have the properties of projection operators, in such a way that the retarded and advanced components of $\Gamma(k, \omega)$ do not interfere with each other when the propagators are manipulated, e.g., when the matrix Γ is inverted.

Specifically, using the property (3.26) of $f^{\pm}(x)$, we find that

$$\varphi^+(x)\,\varphi^+(x) = \varphi^+(x); \tag{5.83}$$

$$\varphi^+(x)\,\varphi^-(x) = \varphi^-(x)\,\varphi^+(x) = 0; \tag{5.84}$$

$$\varphi^-(x)\,\varphi^-(x) = \sigma\varphi^-(x). \tag{5.85}$$

These are precisely the properties of the functions $f^{\pm}(x)$ in the zero temperature limit, where they become step functions with mutually exclusive domains (Eq. 3.29). The functions $\varphi^{\pm}(x)$ share also the property (3.26) of $f^{\pm}(x)$, namely,

$$\varphi^+(x) + \sigma\varphi^-(x) = 1, \tag{5.86}$$

and it is handy further to define the matrix analog of $\alpha(x)$ (Eq. 3.71):

$$\gamma(x) = \varphi^+(x) - \sigma\varphi^-(x) \tag{5.87}$$

$$= \begin{bmatrix} \alpha(x) & 1 + \alpha(x) \\ 1 - \alpha(x) & -\alpha(x) \end{bmatrix}, \tag{5.88}$$

with the property,

$$\gamma(x)\,\gamma(x) = 1, \tag{5.89}$$

like that exhibited by $\alpha(x)$ at zero temperature.

The spectral form of $\Gamma(k, \omega)$ can be read off from that for $G^R(k, \omega)$ and $G^A(k, \omega)$ (Eq. 3.75), together with a form analogous to Eq. (3.70):

$$\Gamma(k, \omega) = \int \varrho(k, \omega') \left(\frac{\varphi^+(\bar{\omega})}{\omega - \omega' + i\eta} + \sigma \frac{\varphi^-(\bar{\omega})}{\omega - \omega' - i\eta} \right) d\omega' \tag{5.90}$$

$$= \mathscr{P} \int \frac{\varrho(k, \omega')}{\omega - \omega'}\, d\omega' - i\pi\gamma(\bar{\omega})\,\varrho(k, \omega). \tag{5.91}$$

[The first term in Eq. (5.91) is understood to be proportional to the unit 2×2 matrix]. In Eq. (5.90), $\varphi^{\pm}(\bar{\omega})$ can be replaced by $\varphi^{\pm}(\bar{\omega}')$, since it is only in the second term of Eq. (5.91), where $\omega' = \omega$, that the argument of $\varphi^{\pm}$

enters. This gives an alternative form, like that of Eq. (3.68), in which the explicit ω dependence is simple:

$$\Gamma(k, \omega) = \int \varrho(k, \omega') \left(\frac{\varphi^+(\bar{\omega}')}{\omega - \omega' + i\eta} + \sigma \frac{\varphi^-(\bar{\omega}')}{\omega - \omega' - i\eta} \right) d\omega'. \tag{5.92}$$

The unperturbed propagator $\Gamma_0(k, \omega)$ is easily obtained from the spectral form (5.90) by taking $\varrho_0(k, \omega) = \delta(\omega - \varepsilon_k)$:

$$\Gamma_0(k, \omega) = \frac{\varphi^+(\bar{\omega})}{\omega - \varepsilon_k + i\eta} + \sigma \frac{\varphi^-(\bar{\omega})}{\omega - \varepsilon_k - i\eta} \tag{5.93}$$

$$= (\omega - \varepsilon_k)^{-1} - i\pi\gamma(\bar{\omega})\,\delta(\omega - \varepsilon_k). \tag{5.94}$$

The derivation of the prescription for the perturbation expansion of $\Gamma(k, \omega)$ follows exactly the same lines as that for $G^c(k, \omega)$ (Eq. 4.50) in the ground state theory. The symbolic prescription is in exact analogy to Eq. (4.50), and is

$$\Gamma_{ab}(k, \omega)\,\delta(\omega - \omega_0) = \sum^{(L)}(-1)^L \sum_{\{k_r\}} \sum_{\{a_i\}} \int \prod d\omega_r$$

$$\times \prod_i [(i/2\pi)\,\sigma_{a_i} v\delta(\omega_r + \omega_r - \omega_{r'} - \omega_{s'})] \prod_r \Gamma_0(k_r, \omega_r). \tag{5.95}$$

The explanation is the same, except that a line which begins and ends at the same vertex now corresponds to a factor $\Gamma_{0aa}(k, \omega)\, e^{i\sigma_a \epsilon \omega}$, where ϵ is a positive infinitesimal. The prescription for $\Sigma(k, \omega)$ takes the identical form, the only difference being the limitation to proper self-energy graphs.

The Fourier transform of the expansion (5.78) for Γ in terms of Σ takes the same form as that (Eq. 4.69) for G^c in terms of M^c at $T = 0$:

$$\Gamma(k, \omega) = \Gamma_0(k, \omega) + \Gamma_0 \Sigma \Gamma_0 + \Gamma_0 \Sigma \Gamma_0 \Sigma \Gamma_0 + \cdots, \tag{5.96}$$

where the arguments are (k, ω) in every factor, and where again matrix multiplication is understood. This sort of matrix geometric series can be summed as readily as the conventional geometric series, and yields

$$\Gamma = (\Gamma_0^{-1} - \Sigma)^{-1}, \tag{5.97}$$

where the arguments (k, ω) are again understood. [This can be demonstrated by multiplying both sides of Eq. (5.96) on the left (or on the right) by $\Gamma_0^{-1} - \Sigma$, and checking the cancellation term by term of all terms but the first.] But it is clear from Eq. (5.94) that

$$\Gamma_0^{-1}(k, \omega) = \omega - \varepsilon_k, \tag{5.98}$$

since

$$(\omega - \varepsilon_k)\,\delta\,(\omega - \varepsilon_k) = 0; \tag{5.99}$$

we therefore find that

$$\Gamma(k, \omega) = [\omega - \varepsilon_k - \Sigma(k, \omega)]^{-1}, \tag{5.100}$$

the exact analog of Eq. (4.70) for $G^c(k, \omega)$ at $T = 0$. The matrix Σ is in general nonhermitian, so that Γ has no singularity, but if one or both of the eigenvalues of the matrix Σ are real, further specification will be needed to make the inverse unambiguous.

Now because of the projection-operator properties (Eqs. 5.83–86) of φ^+ and $\sigma\varphi^-$, we see that

$$\Gamma^{-1}(k, \omega) = \varphi^+(\bar{\omega})\, G^R(k, \omega)^{-1} + \sigma\varphi^-(\bar{\omega})\, G^A(k, \omega)^{-1}, \tag{5.101}$$

as can be confirmed by direct multiplication with Eq. (5.81). However, $G^R(k, \omega)$ and $G^A(k, \omega)$ are the boundary values at real z of the analytic propagator $G(k, z)$ (Eq. 3.81), whose inverse we know (Eq. 5.58):

$$G^{-1}(k, z) = z - \varepsilon_k - M(k, z). \tag{5.102}$$

If we define retarded and advanced self-energy functions $M^{R,A}(k, \omega)$ as the boundary values of $M(k, z)$, in the same way as Eq. (3.81),

$$M^{R,A}(k, \omega) = M(k, \omega \pm i\eta), \tag{5.103}$$

then

$$G^{R,A}(k, \omega)^{-1} = \omega - \varepsilon_k - M^{R,A}(k, \omega), \tag{5.104}$$

and Eq. (5.101) gives

$$\Gamma^{-1}(k, \omega) = \omega - \varepsilon_k - \varphi^+(\bar{\omega})\, M^R(k, \omega) - \sigma\varphi^-(\bar{\omega})\, M^A(k, \omega). \tag{5.105}$$

Here we have used Eq. (5.86) for the terms proportional to $(\omega - \varepsilon_k)$. Comparing Eqs. (5.100) and (5.105), we see finally that

$$\Sigma(k, \omega) = \varphi^+(\bar{\omega})\, M^R(k, \omega) + \sigma\varphi^-(\bar{\omega})\, M^A(k, \omega), \tag{5.106}$$

a relation of the same form as that for $\Gamma(k, \omega)$ (Eq. 5.81). This relation could have been derived, like the corresponding one for $\Gamma(k, \omega)$, from the definitions (Eqs. 5.63, 76) of $M(k, z)$ and $\Sigma(k, t)$ and a periodicity relation for $M(k, t)$, exactly like that (Eq. 5.31) satisfied by $G(k, t)$, which can be proved term by term directly from the rules (Eq. 5.33) for calculating $M(k, t)$.

The matrix Γ will display a pole on the real axis in ω if the determinant of Γ^{-1} (Eq. 5.100) vanishes. Because of the properties (5.83–86) of the singular matrices $\varphi^\pm$, the eigenvalues of Σ can be seen to be just M^R and M^A, which are complex conjugates of each other (Eqs. 5.104, 3.75). Thus Γ can have a pole on the real axis (it is not defined for complex ω) only where M^R and, hence, M^A are real. In this case G^R and G^A have poles at the same point, which (from Eq. 3.75) corresponds to a delta function term in the spectral function $\varrho(k, \omega)$. The pole in G^R must thus have a term $+i\eta$ in the denominator, while that in G^A must have $-i\eta$. This, through Eq. (5.81), determines the character of the singularity in $\Gamma(k, \omega)$ in this case: it is, not surprisingly, of the same kind as that in $\Gamma_0(k, \omega)$ (Eq. 5.93).

In summary, this expansion procedure is as follows: (1) enumerate proper self-energy graphs and calculate $\Sigma(k, \omega)$ using the prescription of Eq. (5.95); (2) for any set of graphs (proof is not given here) this gives $\Sigma(k, \omega)$ in the form of Eq. (5.106), from which $M^R(k, \omega)$ [and its complex conjugate $M^A(k, \omega)$] can be identified; (3) Eq. (5.104) then gives $G^R(k, \omega)$, whose imaginary part, from Eqs. (3.75, 65), is just $-\pi\varrho(k, \omega)$; from the spectral function the thermodynamic functions are determined, through Eqs. (3.92, 97) together with any other propagators desired.

As has been mentioned, the very singular nature, at $\omega = \varepsilon_k$, of the different terms in the expansion (5.96) of Γ in terms of Γ_0 and Σ makes it desirable to use the summed form (5.97) for Γ wherever self-energy subgraphs appear. That is, in calculating $\Sigma(k, \omega)$, we should restrict ourselves to irreducible graphs only, and use Γ, or some approximation to it, rather than Γ_0 for each line of the graph.

We illustrate this procedure first by calculating the Hartree–Fock-type graphs of Figure 9. The prescription gives, for the two terms,

$$\Sigma_{(9)ab}(k, \omega) = -\sigma\delta_{ab} \sum_{k'} \int d\omega_1\, (i/2\pi)\, \sigma_a v\,(k, k_1; k, k_1)\, \Gamma_{aa}(k_1, \omega_1)\, e^{i\sigma_a\varepsilon\omega_1}$$

$$+ \delta_{ab} \sum_{k_1} \int d\omega_1\, (i/2\pi)\, \sigma_a v\,(k, k_1; k_1, k)\, \Gamma_{aa}(k_1, \omega_1)\, e^{i\sigma_a\varepsilon\omega_1}, \qquad (5.107)$$

where $(-1)^L$ is replaced by $-\sigma$ in the first term, in order to give -1 for fermions and $+1$ for bosons. The ω_1 integrations can be done if we use for $\Gamma(k_1, \omega_1)$ the form (5.92), in which the explicit ω-dependence is simple. Since

$$\int (\omega_1 - \omega' \pm i\eta)^{-1} e^{\pm i\varepsilon\omega}\, d\omega_1 = 0, \qquad (5.108)$$

$$\int (\omega_1 - \omega' \mp i\eta)\, e^{\pm i\varepsilon\omega_1}\, d\omega_1 = \pm 2\pi i\, e^{\pm i\varepsilon\omega'} \to \pm 2\pi i \quad (\varepsilon \to 0), \qquad (5.109)$$

we find, using Eqs. (5.92, 82), that

$$\begin{aligned}\int \Gamma_{aa}(k_1, \omega_1)\, e^{i\sigma_a\varepsilon\omega_1}\, d\omega_1 &= \int \varrho\,(k_1, \omega')\, d\omega'\, [-2\pi i\delta_{a,2}\varphi^{+}_{22}(\bar{\omega}') \\ &\quad + 2\pi i\sigma\delta_{a,1}\varphi^{-}{}_{11}(\bar{\omega}')] \\ &= 2\pi i\sigma_a\sigma \int f^{-}(\bar{\omega}')\, \varrho\,(k_1, \omega')\, d\omega' \\ &= 2\pi i\sigma_a\sigma\, \langle \mathbf{n}_{k_1} \rangle,\end{aligned} \qquad (5.110)$$

where $\langle \mathbf{n}_k \rangle$ is the average number of particles of momentum k (Eq. 3.92) for the true system, and the factor $\sigma_a f^-(\bar{\omega}')$ simply summarizes the components of $\varphi^\pm$ from the preceding line. Substituting in Eq. (5.107) now, we get

$$\Sigma_{(9)ab}(k, \omega) = \delta_{ab} \sum_{k_1} [v(k, k_1; k, k_1) - \sigma v(k, k_1; k_1, k] \quad \langle \mathbf{n}_k \rangle \tag{5.111}$$

$$= \delta_{ab} V(k), \tag{5.112}$$

where $V(k)$ is the Hartree–Fock effective potential:

$$V(k) = \sum_{k_1} [v(k, k_1; k, k_1) - \sigma v(k, k_1; k_1, k)] \quad \langle \mathbf{n}_k \rangle. \tag{5.113}$$

This is trivially cast in the form (5.106) by using Eq. (5.86):

$$\Sigma_{(9)}(k, \omega) = V(k) [\varphi^+(\bar{\omega}) + \sigma\varphi^-(\bar{\omega})], \tag{5.114}$$

so that, if this is the only term in $\Sigma(k, \omega)$ that we keep, we have

$$M^R(k, \omega) = M^A(k, \omega) = V(k). \tag{5.115}$$

In this case (5.104),

$$G^{R,A}(k, \omega) = [\omega - \varepsilon_k - V(k) \pm i\eta]^{-1}, \tag{5.116}$$

and hence (3.78),

$$\varrho(k, \omega) = \delta[\omega - \varepsilon_k - V(k)]; \tag{5.117}$$

we thus have a free-particle type of behavior with single-particle energy $\varepsilon_k + V(k)$. Expression (3.97) for the total energy is seen to give the correct Hartree–Fock energy:

$$E = \sum_k [\varepsilon_k + \tfrac{1}{2}V(k)] \quad \langle \mathbf{n}_k \rangle. \tag{5.118}$$

Another more complicated example would be the graph of Figure 11, for which our rules give the following contribution:

$$\Sigma_{(11)ab}(k, \omega) = -\sigma \sum_{k_1 k_2} \sum_c \int d\omega_1 \, d\omega_2 \, d\Omega \, (i/2\pi)^3$$

$$\times v(k, K - k; k_1, K - k_1) \, v(k_1, K - k_1; k_2, K - k_2) \, v(k_2, K - k_2; k, K - k)$$

$$\times \Gamma_{ac}(k_1, \omega_1) \Gamma_{ac}(K - k_1, \Omega - \omega_1) \Gamma_{cb}(k_2, \omega_2) \Gamma_{cb}(K - k_2, \Omega - \omega_2)$$

$$\times \Gamma_{ba}(K - k, \Omega - \omega). \tag{5.119}$$

This could be calculated directly if the propagators Γ were given numerically, say; if the spectral form (5.92) is used for these factors Γ, on the other hand, then the ω integrations of Eq. (5.119) can be done explicitly, leaving an expression essentially identical in form to that given by the prescription (5.71) summed over the six graphs of Figure 12.

5.7 *Many-boson system with condensed phase*

The use of the grand canonical ensemble for a boson system with a condensed phase, i.e. a macroscopically occupied zero-momentum state, involves some rather subtle considerations, since the chemical potential differs from the lowest single-particle energy by an amount proportional to $1/V$. This means that the statistical functions $f^{\pm}$ are being evaluated very close to the singularity, and the correct separation of the volume dependence becomes complicated. In the absence of the condensed phase this problem does not occur, and it is reasonable to suppose, if the interaction is weak and the properties of the actual system are closely similar to those of the unperturbed system at the same temperature, that the perturbation expansion using the grand ensemble method is applicable. If the unperturbed system has a condensed phase, while the actual system does not, for the same values of β and μ, then one can in principle start with a different value of μ for the unperturbed system, chosen to give it a low enough density not to have a condensed phase, and then introduce a term $\mathbf{N}\delta\mu$ into the perturbing Hamiltonian to give the perturbed system the right density. This introduces only formal manipulations, equivalent in the end to ignoring the problem altogether; perhaps this just has to do with the convergence or nonconvergence of the self-energy series (5.34), since the formal result is the same whether the term $\mathbf{N}\delta\mu$ is included in $\bar{H}_0$ or in H'.

If the actual system does have a condensed phase, on the other hand, we can conveniently use the canonical ensemble (fixed N), and use the method applied in the last chapter to the ground state to remove the zero-momentum state from the dynamical problem. This results again in an equivalent artificial problem in which the *grand* ensemble is used for a system without a zero-momentum state and without a condensed phase, but with terms in the Hamiltonian corresponding to production and absorption of particles. The proof of validity, omitted here, is by a modification and extension of the method of Kromminga and Bolsterli [41].

We follow a procedure essentially the same as for the ground state problem. We start again with the modified Hamiltonian $\bar{H}$ (Eq. 4.72), which does not alter the dynamics since N is being held fixed. Our use of the letter μ in defining $\bar{H}$ anticipates the result that μ is equal to the chemical potential for the true system, but for now it is to be regarded as a parameter, to be adjusted to make the Helmholtz free energy insensitive to changes in N_0. In $\bar{H}$ we can, as before, simply replace the operator a_0 by the c-number $N_0^{1/2}$, which leaves us with a system without particle number conservation; the canonical ensemble therefore reduces to the grand ensemble, with chemical potential equal to zero for this artificial system (μ, remember, is to be regarded as merely an adjustable parameter). Our perturbation expansion methods are thus applicable, yielding a prescription for propagators which differ, like those of section 4.6, by factors $e^{\pm i\mu t}$ from those defined for normal systems. The parameters μ and N_0 are now determined by the conditions

$$\partial \bar{A}\,(\mu, N_0)/\partial N_0 = 0, \tag{5.120}$$

and

$$\mathcal{N}\,(\mu, N_0) = N. \tag{5.121}$$

Here $\bar{A}\,(\mu, N_0)$ is the Helmholtz free energy for the artificial system,

$$\bar{A}\,(\mu, N_0) = -\beta^{-1} \ln \operatorname{Tr} e^{-\beta \bar{H}(\mu, N_0)}, \tag{5.122}$$

which reduces to $\bar{E}\,(\mu, N_0)$ at zero temperature. We see by differentiating this expression that

$$\partial \bar{A}\,(\mu, N_0)/\partial \mu = \langle \partial \bar{H}/\partial \mu \rangle = -N_0 - \langle \mathbf{N}' \rangle = -N, \tag{5.123}$$

provided the proper value $N_0(\mu)$ is inserted for N_0, in exact analogy to Eq. (4.86). Combining this with (5.120), we see [cp. (4.87–89)] that the true function $\bar{A}$ satisfies

$$\partial \bar{A}\,(\mu, T, V)/\partial \mu = \partial \bar{A}\,[\mu, N_0(\mu)]/\partial \mu = -N. \tag{5.124}$$

Since $\bar{A}$ turns out to be related to the true Helmholtz function A by

$$\bar{A} = A - \mu N, \tag{5.125}$$

we find that

$$\partial A\,(N, T, V) = \mu. \tag{5.126}$$

Thus μ is the true chemical potential, or Gibbs function per particle. For this reason [see Eqs. (4.90, 91)],

$$\bar{A} \equiv -PV, \tag{5.127}$$

where P is the pressure, just as one would expect if $-\beta\bar{A}$ had been defined as the log of the grand partition function for the true system [see Eq. (5.122)].

We need $\bar{A}\,(\mu, N_0)$ for arbitrary μ and N_0, in order to use the condition (5.120). Now the energy $\bar{E}\,(\mu, N_0)$ of the artificial system can no longer be calculated from Eq. (4.74), which is not valid for finite temperatures, and while we could directly obtain $\bar{E}$ as the expectation value of $\bar{H}$, and deduce $\bar{A}$ from that, it is probably more straightforward to calculate $\ln \mathcal{Q}$, which equals $-\beta\bar{A}$, from Eq. (3.103), which is valid here even though μ and No are independent parameters. The particle number $\mathcal{N}(\mu, N_0)$ is still directly obtainable from the single-particle propagator.

We can now apply directly any of the techniques developed in this section, using the modified Hamiltonian $\bar{H}(\mu, N_0)$ (Eq. 4.73). The enumeration of Feynman graphs is the same as for the ground-state case, with the basic vertices of Figure 6, with the difference that graphs containing lines directed backward in time, like Figure 7(e), can now give nonzero contributions. In each of the various prescriptions (5.27, 60, 69, 95) that we have worked out, there were factors associated with particle lines which must now be omitted, along with the propagators themselves, in order to compensate for the absence of operators a_0 and $a_0^\dagger$ in $\bar{H}(\mu, N_0)$. Before consolidation of factors $i, 2\pi$, β, and σ_a, these were as follows: for the time contour prescription (5.27), a factor i; for the imaginary energy prescription (5.60), a factor $-\beta^{-1}$; for the analytic transform prescription (5.69), a factor $f^{\pm}(\bar{\varepsilon}_n)$; and for the real energy prescription (5.95), a factor $i(2\pi)^{-1}\sigma_b$ [associated with propagator $\Gamma_{0ab}(k, \omega)$]. Recalling (Eq. 4.77) that the energy $\bar{\omega}$, $\bar{\varepsilon}_n$, or $\bar{z}$ is to be associated with particle lines, we give as an example the contribution of Figure 7(e) to M, using the various prescriptions mentioned above, for the case that $k \neq 0$:

$$M_{7(e)}(k, t) = iN_0 \sum_{k_1}{}' v(k, k_1; k+k_1, 0)\, v(k+k_1, 0; k, k_1)\, G_0(k_1, -t) \times G_0(k+k_1, t) \quad (t \in C); \tag{5.128}$$

$$M_{7(e)}(k, z_\nu) = -\beta^{-1} N_0 \sum_{k_1}{}' \sum_{\nu_1} v(k, k_1; k+k_1, 0)\, v(k+k_1, 0; k, k_1) \times G_0(k_1, z_{\nu_1})\, G_0(k+k_1, z_{\nu+\nu_1}); \tag{5.129}$$

$$M_{7(e)}(k, z) = N_0 \sum_{k_1}{}' v(k, k_1; k+k_1, 0)\, v(k+k_1, 0; k, k_1) \times (z + \varepsilon_{k_1} - \varepsilon_{k+k_1-\mu})^{-1} [f^+(\bar{\varepsilon}_{k+k_1}) f^-(\bar{\varepsilon}_{k_1}) - f^-(\bar{\varepsilon}_{k+k_1}) f^+(\bar{\varepsilon}_{k_1})]; \tag{5.130}$$

$$\Sigma_{7(e)ab}(k, \omega) = i(2\pi)^{-1} N_0 \sigma_a \sum_{k_1}{}' \int d\omega_1\, v(k, k_1; k+k_1, 0) \times v(k+k_1, 0; k, k_1)\, \Gamma_{0ba}(k_1, \omega_1)\, \Gamma_{0ab}(k+k_1, \omega+\bar{\omega}_1). \tag{5.131}$$

In these expressions, which all contain the same information in different forms, the unperturbed propagators are those of the grand ensemble, with chemical potential zero, but μ appearing as part of the particle energies; the sums over k_1 do not include $k_1 = 0$, though in passing to integrals in the large-volume limit this exclusion may be ignored. From each of the expressions (5.128–131) the analytic self-energy function can be worked out, using Eq. (5.63) (with z replaced by $\bar{z}$) in the first case, and Eqs. (5.106, 103) in the last. We get in each case

$$M_{7(e)}(k, z) = N_0 \sum_{k_1}' v(k, k_1; k + k_1, 0) \times v(k + k_1, 0; k, k_1) \frac{f^-(\bar{\varepsilon}_{k_1}) - f^-(\bar{\varepsilon}_{k+k_1})}{z + \varepsilon_{k_1} - \varepsilon_{k+k_1} - \mu}. \tag{5.132}$$

5.8 Zero-temperature limit

Let us consider briefly the way in which the finite temperature theory goes over, for $T = 0$, into the ground state theory of the preceding section. We need not give special attention to the boson system with condensed phase, since the treatment which follows is applicable to the equivalent artificial problem, and the relations between the true problem and the artificial one are exactly the same at zero and at nonzero temperatures.

We can deal with each of the three prescriptions (5.60, 71, 95) in turn; the first and last yield in the limit the prescription (4.50) which we have already derived, while the rules (5.71) for the analytic function $M(k, z)$ are unaltered except in the replacement of the statistical factors $f^\pm$ by their step function limits (Eq. 3.29). That is, we obtain a prescription, which we could have derived in Chapter 4, for the ground-state analytic self-energy function $M(k, z)$ of the same form as Eq. (5.71):

$$M(k, z) = \sum^{(l)} (-1)^L \sum_{\{k_r\}} \int \prod d\omega_r \prod_r \varrho(k_r, \omega_r) \prod_i v \times \prod_\alpha D_\alpha^{-1} \prod_r^{(1)} \theta(\bar{\omega}_r) \prod_r^{(2)} [-\theta(-\bar{\omega}_r)] \quad (T = 0), \tag{5.133}$$

where again we avoid the anomalous diagram problem by using only irreducible graphs, and the true spectral function, or some approximation to it, for each internal line.

The prescription (5.60), or the corresponding prescription for $M(k, z_\nu)$, involves sums over the discrete values z_ν of the complex variable z (Eq. 5.39). These values are separated by $2\pi i/\beta$, so that as $\beta \to \infty$ they become densely distributed along a contour passing from $\mu - i\infty$ to $\mu + i\infty$. The sums can be replaced by integrals in the limit, since the summand is analytic in the variables z, except at $z = \mu$, where care must be taken. The limiting relation is

$$\sum_{\nu=-\infty}^{\infty} \rightarrow \int_{-\infty}^{\infty} d\nu = \frac{\beta}{2\pi i} \int_{\mu-i\infty}^{\mu+i\infty} dz. \tag{5.134}$$

The contour of integration, which we call C_μ is shown in Figure 16, and, apart from a little trouble with lines beginning and ending at the same vertex, can be distorted to the contour C_F (the "Feynman contour") also shown in Figure 16. The function of z being calculated must of course be analytically continued at the same time.

The use of the contour C_F is equivalent to replacing $G_0(k, z)$ by

$$\theta(\bar{\omega})\, G_0(k, \omega + i\eta) + \theta(-\bar{\omega})\, G_0(k, \omega - i\eta),$$

which, by Eqs. (3.81, 74, 29) is simply $G_0^c(k, \omega)$. When the conservation of energy at the vertices is taken into account [one can again use contour-dependent delta functions similar to that defined in Eq. (5.32), passing to

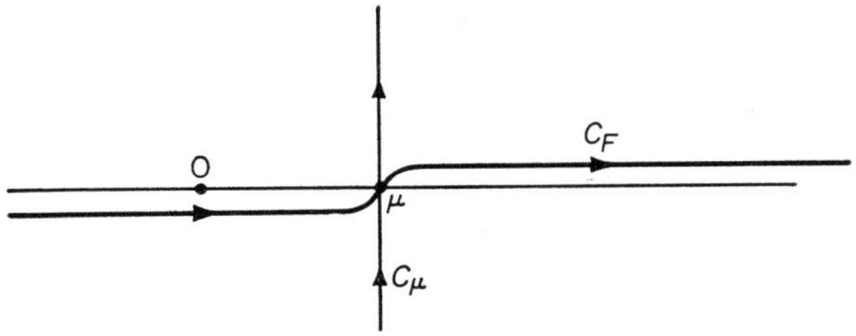

FIGURE 16 Contours of integration C_μ and C_F in the complex plane, for the zero-temperature limit

conventional delta functions for real z], and the factors β and $2\pi i$ are sorted and counted, we find we have precisely the prescription (4.50). For lines which begin and end at the same vertex, the factor $e^{\epsilon z}$ in Eq. (5.60) means that the contour C_μ can be closed in the left half-plane with a semicircle of infinite radius, which becomes a semicircle in the upper half plane when C_μ is distorted to C_F. This corresponds to the factor $e^{i\epsilon\omega}$ to be found in Eq. (4.50).

The analysis for the real-time theory (Eq. 5.95) is quite different, but the result is the same. In this case we find that if the endpoints of a graph are on C_1 (Figure 15), then the contribution of C_2 vanishes when $T \rightarrow 0$, and the T_C-ordered product (Eq. 5.19) reduces to the T-ordered product of Eq. (4.35). That is to say, the contribution of terms in Eq. (5.95) for which some of the a_i are equal to 2 vanishes in the limit. To see this, let us deform a given graph in such a way that all those vertices for which $a = 2$ can be enclosed in an imaginary circle. This part of the graph will be connected to the remainder by equal numbers of lines entering and leaving, all of which connect an $a = 1$ vertex outside the circle to an $a = 2$ vertex inside. Each line leaving the circle corresponds to a factor $\Gamma_{12}(k_r, \omega_r)$ (using the irreducible graph procedure, to be specific), which contains (Eqs. 5.81, 82) the factor $f^-(\bar{\omega}_r)$; and each line entering, to a factor $\Gamma_{21}(k_r, \omega_r)$, which contains the factor $f^+(\bar{\omega}_r)$. Now energy conservation at the vertices ensures that the sum of the arguments ω_r for the lines entering the circle is equal to the sum for the lines leaving; on the other hand, the functions $f^{\pm}(\bar{\omega}_r)$ reduce to step functions (Eq. 3.29) in the zero-temperature limit, so that all the lines leaving the circle must have arguments ω_r which are less than μ, and all the lines entering must have arguments ω_r greater than μ. This contradicts the conservation of energy (the case in which all ω_r's are equal to μ giving no contribution upon integration) and so the graph does not contribute. Since all the vertices must thus have $a = 1$, then, we find that each line is now represented by $\Gamma_{11}(k, \omega)$, which is just $G^c(k, \omega)$ (Eq. 5.80), and we are back to the irreducible graph form of Eq. (4.50) (in which, as was discussed in Chapter 4, only irreducible self-energy graphs are enumerated, and G_0^c is replaced by G^c on the lines of each graph).

Note added in proof (March 1969)

The transformation (5.42) can be reduced to a conventional Fourier transformation (see section 5.4), so that in addition to the general relation (5.54), the functions $K_i(z_\nu)$ also obey the normal sort of folding relation:

$$K_3(z_{\nu_3}) = \sum K_1(z_{\nu_1})\, K_2(z_{\nu_2})\, \delta\,(\nu_3, \nu_1 + \nu_2).$$

Note that, because of (5.52, 53),

$$z_{\nu_3} = z_{\nu_1} + z_{\nu_2} \qquad (\text{for } \nu_3 = \nu_1 + \nu_2).$$

Appendices

APPENDIX A

Summary of properties of propagators

WE SHALL here summarize as concisely as possible the properties of the different propagators and self-energy functions with which we have to deal, and of their Fourier transforms. We give in each case the defining relation, the expression in terms of the basic propagators $G^{\pm}(k, t)$ for the time argument, and in terms of $G^{R,A}(k, \omega)$, or $G(k, z)$ for the energy argument, the spectral representation [the spectral function is given by Eqs. (3.50, 87–89)], the graphical prescription in symbolic form (for two-body forces) in case there is one, the free-particle form, and the domain of definition. The relation between the propagators and self-energy functions is given, though the latter are taken here as defined, not by this relation, but by the specification of graphs to be summed over. Arguments will occasionally be suppressed, where they are the same as in the function being discussed, in the hope of making the structure of an expression clearer. It will be recalled that β is inversely proportional to the temperature and becomes infinite in the ground-state case; the factor (or index) σ indicates the choice of statistics, being $+1$ for fermions and -1 for bosons. We again let $\bar{\omega}$ represent $\omega - \mu$ where it is convenient, and $\bar{\varepsilon}_k$ represent $\varepsilon_k - \mu$. The FD and BE distribution functions $f^{\pm}(x)$ are defined in Eqs. (3.20, 27). In the factor $(-1)^L$ in the various prescriptions, L represents the number of closed fermion loops.

Infinitesimal positive times and energies are represented by ϵ and η respectively. In summing over graphs, we use the following terminology:

$\sum^{(L)}$: sum over linked graphs, i.e., those for which every point is connected to some endpoint by lines and vertices.

$\sum^{(P)}$: sum over proper self-energy graphs, i.e., those for which no two portions are connected by less than two lines. They are necessarily linked.

$\sum^{(I)}$: sum over irreducible self-energy graphs, i.e., those proper graphs containing no self-energy subgraphs.

The vertices of a graph are labeled by $i, j, \ldots$, and the lines, by $r, s, \ldots$; the intermediate states of a time-ordered graph are labeled by α.

Except in one case [the analytic propagator $G(k, z)$] the Fourier transform is defined by

$$G^{(\,)}(k, \omega) = \int_{-\infty}^{\infty} G^{(\,)}(k, t)\, e^{i\omega t}\, dt, \tag{A.1}$$

which implies

$$G^{(\,)}(k, t) = (2\pi)^{-1} \int_{-\infty}^{\infty} G^{(\,)}(k, \omega)\, e^{-i\omega t}\, d\omega. \tag{A.2}$$

The large-volume limit is obtained by the use of Eqs. (3.94) and (2.70).

The basic propagators $G^{\pm}$

$$G^{+}(k, t - t_0) = \langle \mathbf{a}_k(t)\, \mathbf{a}_k^{\dagger}(t_0) \rangle \tag{A.3}$$

$$= iG(k, t - t_0); \tag{A.4}$$

$$G^{-}(k, t - t_0) = \langle \mathbf{a}_k^{\dagger}(t_0)\, \mathbf{a}_k(t) \rangle \tag{A.5}$$

$$= -i\sigma G(k, t - t_0); \tag{A.6}$$

$$G^{\pm}(k, t) = \int_{-\infty}^{\infty} \varrho(k, \omega) f^{\pm}(\bar{\omega})\, e^{-i\omega t}\, d\omega; \tag{A.7}$$

$$G_0^{\pm}(k, t) = f^{\pm}(\bar{\varepsilon}_k)\, e^{-i\varepsilon_k t}; \tag{A.8}$$

$$0 \leqslant \mp \operatorname{Im}(t - t_0) \leqslant \beta. \tag{A.9}$$

$$G^{\pm}(k, \omega) = if^{\pm}(\bar{\omega})(G^R - G^A) \tag{A.10}$$

$$= 2\pi\varrho(k, \omega) f^{\pm}(\bar{\omega}); \tag{A.11}$$

$$G_0^{\pm}(k, \omega) = 2\pi\delta(\omega - \varepsilon_k) f^{\pm}(\bar{\varepsilon}_k); \tag{A.12}$$

Referred to as $G^{>}$ and $G^{<}$ by many authors.

The retarded and advanced propagators $G^{R,A}$

$$G^{R,A}(k, t - t_0) = \mp i\theta[\pm(t - t_0)]\, \langle [\mathbf{a}_k(t), \mathbf{a}_k^{\dagger}(t_0)]_{\sigma} \rangle; \tag{A.13}$$

$$G^{R,A}(k, t) = \mp i\theta(\pm t)(G^{+} + \sigma G^{-}) \tag{A.14}$$

$$= \mp i\theta(\pm t) \int_{-\infty}^{\infty} \varrho(k, \omega)\, e^{-i\omega t}\, d\omega; \tag{A.15}$$

$$G_0^{R,A}(k, t) = \mp i\theta(\pm t)\, e^{-i\varepsilon_k t}; \tag{A.16}$$

t is real.

$$G^{R,A}(k, \omega) = G(k, \omega \pm i\eta) \tag{A.17}$$

$$= \int_{-\infty}^{\infty} \varrho(k, \omega')(\omega - \omega' \pm i\eta)^{-1} d\omega' \tag{A.18}$$

$$= [\omega - \varepsilon_k - M^{R,A}(k, \omega)]^{-1}; \tag{A.19}$$

$$G^{R,A}(k, \omega) = (\omega - \varepsilon_k \pm i\eta)^{-1}; \tag{A.20}$$

ω is real, but G^R can be analytically continued in ω to the entire upper half plane, and G^A, to the lower.

The causal and anticausal propagators G^c, $G^{\sim c}$

$$G^c(k, t - t_0) = -i\langle T[\mathbf{a}_k(t)\,\mathbf{a}_k^\dagger(t_0)]\rangle; \tag{A.21}$$

$$G^c(k, t) = -i\theta(t)\, G^+ + i\sigma\theta(-t)\, G^- \tag{A.22}$$

$$= -i \int \varrho(k, \omega)\, e^{-i\omega t}[\theta(t) f^+(\bar{\omega}) - \sigma\theta(-t) f^-(\bar{\omega})]\, d\omega \tag{A.23}$$

$$= G_0^c(k, t) + \int_{-\infty}^{\infty} dt_1 dt_2\, G_0^c(k, t - t_1)\, M^c(k, t_1 - t_2)\, G^c(k, t_2) \quad (T = 0) \tag{A.24}$$

$$= \sum^{(L)} (-1)^L \sum_{\{k_r\}} \int \prod dt_i \prod_i [iv(r, s; r', s')] \prod_r G_0^c(k_r, t_i - t_j - \varepsilon) \quad (T = 0); \tag{A.25}$$

$$G_0^c(k, t) = -i e^{-i\varepsilon_k t}[\theta(t) f^+(\bar{\varepsilon}_k) - \sigma\theta(-t) f^-(\bar{\varepsilon}_k)]; \tag{A.26}$$

t and t_0 are real.

$$G^c(k, \omega) = f^+(\bar{\omega})\, G^R + \sigma f^-(\bar{\omega})\, G^A \tag{A.27}$$

$$= \int \varrho(k, \omega') \left[\frac{f^+(\bar{\omega}')}{\omega - \omega' + i\eta} + \sigma \frac{f^-(\bar{\omega}')}{\omega - \omega' - i\eta}\right] d\omega' \tag{A.28}$$

$$= G_0^c + G_0^c M^c G^c \quad (T = 0) \tag{A.29}$$

$$= [\omega - \varepsilon_k - M^c(k, \omega)]^{-1} \quad (T = 0); \tag{A.30}$$

$$G^c(k, \omega)\, \delta(\omega - \omega_0) = \sum^{(L)} (-1)^L \sum_{\{k_r\}} \int \prod d\omega_r \prod_i [(i/2\pi)\, v(r, s; r', s') \times \delta(\omega_r + \omega_s - \omega_{r'} - \omega_{s'})] \prod_r [G_0^c(k_r, \omega_r)\, e^{i\omega_r \varepsilon}] \quad (T = 0); \tag{A.31}$$

$$G_0^c(k, \omega) = \frac{f^+(\bar{\omega})}{\omega - \varepsilon_k + i\eta} + \sigma \frac{f^-(\bar{\omega})}{\omega - \varepsilon_k - i\eta} \tag{A.32}$$

$$= [\omega - \varepsilon_k + i\eta\sigma(\bar{\varepsilon}_k)]^{-1} \quad (T = 0); \tag{A.33}$$

ω is real.

$$M^c(k, t - t_0) = \sum^{(P)} (-1)^L \sum_{\{k_r\}} \int \prod_1^{n-2} dt_i \prod_i [iv(r, s; r', s')]$$

$$\times \prod_r G_0^c(k_r, t_i - t_j - \varepsilon) \quad (T = 0) \tag{A.34}$$

$$= \sum^{(I)} (-1)^L \sum_{\{k_r\}} \int \prod_1^{n-2} dt_i \prod_i [iv(r, s; r', s')]$$

$$\times \prod_r G^c(k_r, t_i - t_j - \varepsilon) \quad (T = 0); \tag{A.35}$$

[$\int \prod_1^{n-2} dt_i$ is replaced by $\delta(t - t_0) \int \prod_1^{n-1} dt_i$ for Hartree–Fock-type graphs (Figure 9).]

$M^c(k, \omega)$: Same prescription as for $G^c(k, \omega)$, except that $\sum^{(L)}$ is replaced by $\sum^{(P)}$ (with $\prod_r G_0^c$) or by $\sum^{(I)}$ (with $\prod_r G^c$).

$$G^{\sim c}(k, t - t_0) = i \langle \tilde{T} [\mathbf{a}_k(t)\, \mathbf{a}_k^\dagger(t_0)] \rangle; \tag{A.36}$$

$$G^{\sim c}(k, t) = i\theta(-t)\, G^+ - i\sigma\theta(t)\, G^- \tag{A.37}$$

$$= i \int \varrho(k, \omega)\, e^{-i\omega t} d\omega\, [\theta(-t) f^+(\bar{\omega}) - \sigma\theta(t) f^-(\bar{\omega})]; \tag{A.38}$$

$$G_0^{\sim c}(k, t) = i\, e^{-i\varepsilon_k t} [\theta(-t) f^+(\bar{\varepsilon}_k) - \sigma\theta(t) f^-(\bar{\varepsilon}_k)]; \tag{A.39}$$

t and t_0 are real.

$$G^{\sim c}(k, \omega) = f^+(\bar{\omega})\, G^A + \sigma f^-(\bar{\omega})\, G^R \tag{A.40}$$

$$= \int \varrho(k, \omega') \left[\frac{f^+(\bar{\omega}')}{\omega - \omega' - i\eta} + \sigma \frac{f^-(\bar{\omega}')}{\omega - \omega' + i\eta} \right] d\omega'; \tag{A.41}$$

$$G_0^{\sim c}(k, \omega) = \frac{f^+(\bar{\omega})}{\omega - \varepsilon_k - i\eta} + \sigma \frac{f^-(\bar{\omega})}{\omega - \varepsilon_\kappa + i\eta}; \tag{A.42}$$

ω is real.

The analytic propagator G

$$G(k, t - t_0) = -i \langle T_C[\mathbf{a}_k(t)\, \mathbf{a}_k^\dagger(t_0)]\rangle \tag{A.43}$$

$$= -i\theta_C(t, t_0)\, G^+ + i\sigma\theta_C(t_0, t)\, G^- \tag{A.44}$$

$$= -i \int \varrho(k, \omega)\, e^{-i\omega(t-t_0)} [\theta_C(t, t_0) f^+(\bar{\omega}) - \sigma\theta_C(t_0, t) f^-(\bar{\omega})]\, d\omega \tag{A.45}$$

$$= G_0 + \int_C dt_1\, dt_2\, G_0(k, t - t_1)\, M(k, t_1 - t_2)\, G(k, t_2 - t_0) \tag{A.46}$$

$$= \sum{}^{(L)}(-1)^L \sum_{\{k_r\}} \int_C \prod dt_i \prod_i [iv(r, s; r', s')] \times \prod_r G_0(k_r, t_i - t_j + i\varepsilon); \tag{A.47}$$

$$G_0(k, t - t_0) = -i\, e^{-i\varepsilon_k(t-t_0)}[\theta_C(t, t_0) f^+(\bar{\varepsilon}_k) - \sigma\theta_C(t_0, t) f^-(\bar{\varepsilon}_k)]; \tag{A.48}$$

$$t, t_0 \in C. \tag{A.49}$$

C is a contour running from an arbitrary point τ, in the complex t plane, to the point $\tau - i\beta$, C itself being arbitrary except that successive points have nonincreasing imaginary parts. T_C is the ordering operator along C, analogous to the ordering operator T along the real axis; $\theta_C(t_1, t_2) = 1$ if t_1 is farther along C than t_2, zero otherwise. $G(k, t - t_0)$ thus defines an analytic function of $t - t_0$, independent of the choice of C in the domain

$$-\beta < \mathrm{Im}\,(t - t_0) < \beta, \tag{A.50}$$

with a cut along the real axis, G being equal to $-iG^+$ below the real axis, and $i\sigma G^-$ above. The value to be chosen if the contour chosen permits t to be on the real axis does depend on the choice of C.

$$G(k, z) = \int_{C\pm} G(k, t)\, e^{izt}\, dt \tag{A.51}$$

$$= G^R(z) \quad (\mathrm{Im}\, z > 0) \tag{A.52}$$

$$= G^A(z) \quad (\mathrm{Im}\, z < 0) \tag{A.53}$$

$$= \int_{-\infty}^{\infty} \varrho(k, \omega)\, (z - \omega)^{-1}\, d\omega \tag{A.54}$$

$$= G_0 + G_0 M G \tag{A.55}$$

$$= [z - \varepsilon_k - M(k, z)]^{-1}; \tag{A.56}$$

$$G_0(k, z) = (z - \varepsilon_k)^{-1}; \tag{A.57}$$

$$\mathrm{Im}\, z \neq 0; \tag{A.58}$$

the contours $C_{\pm}$ [Figure 10(a, b)] are to be used for Im $z > 0$ and Im $z < 0$, respectively.

$$M(k, t - t_0) = \sum{}^{(P)}(-1)^L \int_C \prod_1^{n-2} dt_i$$

$$\times \sum_{\{k_r\}} \prod_i [iv(r, s; r', s')] \prod_r G_0(k_r, t_i - t_j + i\varepsilon) \qquad \text{(A.59)}$$

$$= \sum{}^{(I)}(-1)^L \int_C \prod_1^{n-2} dt_i$$

$$\times \sum_{\{k_r\}} \prod_i [iv(r, s; r', s')] \prod_r G(k_r, t_i - t_j + i\varepsilon), \qquad \text{(A.60)}$$

with $\int_C \prod_1^{n-2} dt_i$ replaced by $\delta_C(t - t_0) \int_C \prod_1^{n-1} dt_i$ for Hartree–Fock-type graphs (Figure 9).

$$M(k, z) = \sum{}^{(P)}(-1)^L \sum_{\{k_r\}} \prod_i v(r, s; r', s') \prod_\alpha D_\alpha^{-1}$$

$$\times \prod_r{}^{(1)} f^+(\bar{\varepsilon}_r) \prod_r{}^{(2)} [-\sigma f^-(\bar{\varepsilon}_r)] \qquad \text{(A.61)}$$

$$= \sum{}^{(I)}(-1)^L \sum_{\{k_r\}} \int \prod d\omega_r \prod_r \varrho(k_r, \omega_r) \prod_i v(r, s; r', s')$$

$$\times \prod_\alpha D_\alpha^{-1} \prod_r{}^{(1)} f^+(\bar{\omega}_r) \prod_r{}^{(2)} [-\sigma f^-(\bar{\omega}_r)]. \qquad \text{(A.62)}$$

Here the sum is over time-ordered graphs (all possible orderings of vertices from bottom to top of graph); $\prod_r^{(1)}$ indicates a product over lines directed upward, called particle lines, and $\prod_r^{(2)}$, a product over lines directed downward, called hole lines. For D_α, let z_r be the energy associated with each line (either ε_r or ω_r) and let an additional line be drawn from the final vertex back to the initial vertex, labeled k, z, which is to be counted in the same way as all of the other lines. Then for each pair of successive vertices a horizontal cut is to be made between them, corresponding to one energy denominator; letting $\sum_r^{(1)}$ and $\sum_r^{(2)}$ represent sums over the particle lines cut and hole lines cut, respectively, we take

$$D_\alpha = \sum_r{}^{(1)} z_r - \sum_r{}^{(2)} z_r. \qquad \text{(A.63)}$$

For the special case

$$z = z_\nu = \mu + 2\pi i\nu/\beta, \tag{A.64}$$

$$\nu = \cdots, -\tfrac{3}{2}, -\tfrac{1}{2}, \tfrac{1}{2}, \tfrac{3}{2}, \cdots \quad \text{(FD)} \tag{A.65}$$

$$= \cdots, -1, 0, 1, 2, \cdots \quad \text{(BE)}, \tag{A.66}$$

we can take

$$G(k, z_\nu) = \int_{C_0} G(k, t)\, e^{iz_\nu t}\, dt \tag{A.67}$$

$$= \sum^{(L)} (-1)^L \sum_{\{k_r\}} \sum_{\{\nu_r\}} \prod_i [-\beta^{-1} v(r, s; r', s')\, \delta(\nu_r + \nu_s, \nu_{r'} + \nu_{s'})]$$

$$\times \prod_r [G_0(k_r, z_{\nu_r}) \exp(\varepsilon z_{\nu_r})]. \tag{A.68}$$

C_0 [Figure 10(c)] runs along the imaginary axis from a point τ above the real axis, through zero, to the point $\tau - i\beta$. A convenient and usual choice is $\tau = 0 + i\varepsilon$. The self-energy function $M(k, z_\nu)$ is given by the same prescription as $G(k, z_\nu)$, but with $\sum^{(L)}$ replaced by $\sum^{(P)}$ (with $\prod_r G_0$) or by $\sum^{(P)}$ (with $\prod_r G$).

The real-time matrix propagator Γ

Defined with respect to the two real portions C_1 and C_2 of the contour C_R (Figure 15).

$$\Gamma_{ab}(k, t - t_0) = G(k, t - t_0)\, \sigma_b \quad (t \in C_a;\ t_0 \in C_b;\ \sigma_1 = 1;\ \sigma_2 = -1) \tag{A.69}$$

$$= \begin{bmatrix} G^c & -i\sigma G^- \\ -iG^+ & G^{\sim c} \end{bmatrix}_{ab}; \tag{A.70}$$

$$\Gamma_{ab}(k, t) = -i\theta(t) \begin{bmatrix} G^+ \; \sigma G^- \\ G^+ \; \sigma G^- \end{bmatrix}_{ab} + i\sigma\theta(-t) \begin{bmatrix} G^- & -G^- \\ -\sigma G^+ & \sigma G^+ \end{bmatrix}_{ab} \tag{A.71}$$

$$= -i \int \varrho(k, \omega)\, e^{-i\omega t}\, [\theta(t)\, \varphi^+_{ab}(\bar\omega) - \sigma\theta(-t)\, \varphi^-_{ab}(\bar\omega)]\, d\omega \tag{A.72}$$

$$= \Gamma_{0ab} + \sum_{c,d} \int_{-\infty}^{\infty} dt_1\, dt_2\, \Gamma_{0ac}(k, t - t_1)\, \Sigma_{cd}(k, t_1 - t_2)\, \Gamma_{db}(k, t_2) \tag{A.73}$$

$$= \sum^{(L)} (-1)^L \sum_{\{a_i\}} \sum_{\{k_r\}} \int_{-\infty}^{\infty} \prod dt_i \prod_i [i\sigma_{a_i} v(r, s; r', s')]$$

$$\times \prod_r \Gamma_{0a_ia_j}(k_r, t_i - t_j - \sigma_{a_i}\varepsilon); \tag{A.74}$$

$$\Gamma_{0ab}(k, t) = -i\, e^{-\varepsilon_k t}[\theta(t)\, \varphi^{+}_{ab}(\bar{\varepsilon}_k) - \sigma\theta(-t)\, \varphi^{-}_{ab}(\bar{\varepsilon}_k)]; \tag{A.75}$$

t and t_0 are real.

$$\varphi^{+}(\bar{\omega}) = \begin{bmatrix} f^{+} & \sigma f^{-} \\ f^{+} & \sigma f^{-} \end{bmatrix}; \tag{A.76}$$

$$\varphi^{-}(\bar{\omega}) = \begin{bmatrix} f^{-} & -f^{-} \\ -\sigma f^{+} & \sigma f^{+} \end{bmatrix}. \tag{A.77}$$

$$\Gamma(k, \omega) = \int_{-\infty}^{\infty} e^{i\omega t}\, \Gamma(k, t)\, dt \tag{A.78}$$

$$= \begin{bmatrix} G^{c} & -i\sigma G^{-} \\ -iG^{+} & G^{\sim c} \end{bmatrix} \tag{A.79}$$

$$= G^{R}\varphi^{+}(\bar{\omega}) + \sigma G^{A}\varphi^{-}(\bar{\omega}) \tag{A.80}$$

$$= \int \varrho(k, \omega') \left[\frac{\varphi^{+}(\bar{\omega}')}{\omega - \omega' + i\eta} + \sigma \frac{\varphi^{-}(\bar{\omega}')}{\omega - \omega' - i\eta} \right] d\omega' \tag{A.81}$$

$$= \Gamma_0 + \Gamma_0 \Sigma \Gamma \quad \text{(matrix multiplication)} \tag{A.82}$$

$$= (\omega - \varepsilon_k - \Sigma)^{-1}; \tag{A.83}$$

$$\Gamma_{ab}(k, \omega)\, \delta(\omega - \omega_0) = \sum{}^{(L)} (-1)^{L} \sum_{\{a_i\}} \sum_{\{k_r\}} \int_{-\infty}^{\infty} \prod d\omega_r$$
$$\times \prod_{i} [(i/2\pi)\, \sigma_{a_i} v(r, s; r', s')\, \delta(\omega_r + \omega_s - \omega_{r'} - \omega_{s'})]$$
$$\times \prod_{r} [\Gamma_{0a_i a_j}(k_r, \omega_r) \exp(i\sigma_{a_i} \varepsilon \omega)]; \tag{A.84}$$

$$\Gamma_0(k, \omega) = \frac{\varphi^{+}(\bar{\omega})}{\omega - \varepsilon_k + i\eta} + \sigma \frac{\varphi^{-}(\bar{\omega})}{\omega - \varepsilon_k - i\eta}; \tag{A.85}$$

ω is real.

$$\Sigma_{ab}(k, t - t_0) = \sum{}^{(P)} (-1)^{L} \int \prod_{1}^{n-2} dt_i \sum_{\{a_i\}} \sum_{\{k_r\}} \prod_{i} [i\sigma_{a_i} v(r, s; r', s')]$$
$$\times \prod_{r} \Gamma_{0a_i a_j}(k_r, t_i - t_j - \sigma_{a_i}\varepsilon) \tag{A.86}$$

$$= \sum{}^{(I)} (-1)^{L} \int \prod_{1}^{n-2} dt_i \sum_{\{a_i\}} \sum_{\{k_r\}} \prod_{i} [i\sigma_{a_i} v(r, s; r', s')]$$
$$\times \prod_{r} \Gamma_{a_i a_j}(k_r, t_i - t_j - \sigma_{a_i}\varepsilon); \tag{A.87}$$

[$\int \prod_{1}^{n-2} dt_i$ is replaced by $\delta(t - t_0) \int \prod_{1}^{n-1} dt_i$ for Hartree–Fock-type graphs (Figure 9).]

$\Sigma_{ab}(k, \omega)$: same prescription as for $\Gamma_{ab}(k, \omega)$, but with $\sum^{(L)}$ replaced by $\sum^{(P)}$ (with $\prod_r \Gamma_0$) or by $\sum^{(I)}$ (with $\prod_r \Gamma$).

APPENDIX B

Proof of Wick's theorem

WE HERE prove two forms of Wick's theorem, of special usefulness in the many-body problem. The first, which deals with the ground-state expectation value of a T-ordered product is a special case of the original theorem [50], while the second [26, 54, 30, 59], an extension to finite temperature of the first, deals with statistical expectation values in the grand ensemble.

The ground-state form

Let there be $2p$ boson or fermion operators b_r $(r = 1, 2, \ldots, 2p)$, in the interaction representation, and for each r let b_r be in general a linear combination of creation and annihilation operators associated with the time t_r. Let expectation values $\langle (A) \rangle_0$, for arbitrary A, be taken with respect to an independent particle vacuum-like state Φ_0, by which we mean either the actual vacuum state for independent bosons, or the ground state for N independent fermions; the property we need is that b_r can be expressed as the sum of two parts, corresponding to annihilation and creation of excitations above the ground state:

$$b_r = b_r^{(1)} + b_r^{(2)\dagger}, \tag{B.1}$$

with

$$b_r^{(1)}\Phi_0 = 0; \tag{B.2}$$

$$b_r^{(2)}\Phi_0 = 0. \tag{B.3}$$

Then the theorem is that

$$\langle T(\prod_1^{2p} b_r) \rangle_0 = \sum_{\text{a.p.p.}} (-1)^P \prod \langle T(b_r b_s) \rangle_0, \tag{B.4}$$

where the sum is over all possible ways of picking p pairs among the $2p$ operators b_r, and P has the parity of that permutation of fermion operators which is involved in rearranging the pairs in the order in which they are written on the right side, starting from the order on the left side of the equation.

We prove this by induction; it is trivially true for the case $p = 1$, and supposing it to be true for the case $p - 1$, we try to prove it for the case p. Let c_r be the same operators arranged in descending time order, so that if t_q is the latest of the times t_r, then

$$c_1 \equiv b_q, \tag{B.5}$$

and

$$T\left(\prod_1^{2p} b_r\right) = (-1)^{P_1} c_1 \prod_1^{2p} c_r \tag{B.6}$$

$$= (-1)^{P_1} c_1 \prod_2^{2p} c_r, \tag{B.7}$$

where P_1 has the parity of the permutation of fermion operators induced by T. So

$$\left\langle T\left(\prod_1^{2p} b_r\right)\right\rangle_0 = (-1)^{P_1} \left\langle c_1 \prod_2^{2p} c_r \right\rangle_0 \tag{B.8}$$

$$= (-1)^{P_1} \left\langle c_1^{(1)} \prod_2^{2p} c_r \right\rangle_0, \tag{B.9}$$

since the part $c_1^{(2)\dagger}$ (Eq. B.3) would give zero contribution. We now bring $c_1^{(1)}$ through to the right of the expression in brackets, by taking successive commutators (BE) or anticommutators (FD), and use Eq. (B.2) to drop the last term:

$$\left\langle T\left(\prod_1^{2p} b_r\right)\right\rangle_0 = (-1)^{P_1} \sum_{r=2}^{2p} (-\sigma)^{r-1} [c_1^{(1)}, c_r]_\sigma \langle \textstyle\prod' c_s \rangle_0, \tag{B.10}$$

where $\prod'$ is a product over the $2p - 2$ remaining factors. Because the commutator (or anticommutator) here is a c-number, it can be trivially replaced by its expectation value, and the properties (B.2, 3) used to get the following:

$$\begin{aligned} [c_1^{(1)}, c_r]_\sigma &= \langle [c_1^{(1)}, c_r]_\sigma \rangle_0 \\ &= \langle c_1^{(1)} c_r \rangle_0 \\ &= \langle c_1 c_r \rangle_0 \\ &= \langle T(c_1 c_r) \rangle_0 . \end{aligned} \tag{B.11}$$

The remaining factor $\langle\prod' c_s\rangle_0$ is itself a time-ordered product with $2(p-1)$ factors, so that

$$\left\langle T\left(\prod_1^{2p} b_r\right)\right\rangle_0 = (-1)^{P_1} \sum_{r=2}^{2p} (-\sigma)^{r-1} \langle T(c_1 c_r)\rangle_0 \langle T(\textstyle\prod' c_s)\rangle_0. \quad \text{(B.12)}$$

The theorem is valid by assumption for the last factor on the right, which can therefore be expressed as a sum of products of pair expectation values. This, together with the sum over r, enumerates all possible pairings of the original $2p$ factors, and if the sign is right in each term, the theorem is proved. Now at each stage of the proof the changes in sign followed exactly the changes in order of writing fermion operators, and this will continue to be true if the process is continued until the whole product is broken down into pairs; since this is the rule given in the statement of the theorem, the sign will be correct, and the proof is complete.

The finite-temperature form

The statement of the thermodynamic Wick's theorem is just the same as for the ground-state form, except that the expectation value is now taken in the grand ensemble, again for noninteracting particles. Note that the proof of the theorem does not depend on the fact that the t_r's are times—they can be any ordered parameters on which the T-operator can act.

We start by looking at the expectation value of the commutator, or anticommutator, of b_q, say, with any operator, using the cyclic property of the trace:

$$\langle [b_q, A]_\sigma \rangle_0 = \mathcal{Q}_0^{-1} \operatorname{Tr}[e^{-\beta\bar{H}_0}(b_q A + \sigma A b_q)] \quad \text{(B.13)}$$

$$= \mathcal{Q}_0^{-1} \operatorname{Tr}[(e^{-\beta\bar{H}_0} b_q + \sigma b_q e^{-\beta\bar{H}_0}) A]$$

$$= \mathcal{Q}_0^{-1} \operatorname{Tr}[(e^{-\beta\bar{H}_0} b_q + \sigma e^{-\beta(\bar{H}_0 \pm \bar{\varepsilon}_q)} b_q) A]$$

$$= (1 + \sigma e^{\mp\bar{\beta}\varepsilon_q}) \langle b_q A\rangle_0. \quad \text{(B.14)}$$

We have here assumed that b_q is either an annihilation operator (upper sign) or a creation operator (lower sign), and will generalize later to the linear combination. Note that the factor $(1 + \sigma e^{\mp\bar{\beta}\varepsilon_q})$ does not depend on the nature of A. This permits us to express the expectation value $\langle b_q A\rangle_0$ in terms of $\langle [b_q, A]\rangle_0$, and vice versa, for arbitrary A, and hence to follow a line of reasoning analogous to that which starts with Eq. (B.8) and the identification $c_1 \equiv b_q$:

$$\left\langle T\left(\prod_1^{2p} b_r\right)\right\rangle_0 = (-1)^{P_1}\left\langle c_1 \prod_2^{2p} c_r\right\rangle_0$$

$$= (-1)^{P_1}(1 + \sigma\, e^{\mp\bar{\beta}\varepsilon_q})^{-1}\left\langle\left[c_1, \prod_2^{2p} c_r\right]_\sigma\right\rangle_0$$

$$= (-1)^{P_1}(1 + \sigma\, e^{\mp\bar{\beta}\varepsilon_q})^{-1}\sum_{r=2}^{2p}(-\sigma)^{r-1}[c_1, c_r]_\sigma \langle\textstyle\prod' c_s\rangle_0$$

$$= (-1)^{P_1}(1 + \sigma\, e^{\mp\bar{\beta}\varepsilon_q})^{-1}\sum_{r=2}^{2p}(-\sigma)^{r-1}\langle[c_1, c_r]_\sigma\rangle_0 \langle\textstyle\prod' c_s\rangle_0$$

$$= (-1)^{P_1}\sum_{r=2}^{2p}(-\sigma)^{r-1}\langle c_1 c_r\rangle_0 \langle\textstyle\prod' c_s\rangle_0$$

$$= (-1)^{P_1}\sum_{r=2}^{2p}(-\sigma)^{r-1}\langle T(c_1 c_r)\rangle_0 \langle T(\textstyle\prod' c_s)\rangle_0 . \tag{B.15}$$

(The introduction of T-ordering operators in the last step is trivially valid, since the operators were already in this order.) This result is valid whether c_1 is a creation or an annihilation operator, and can thus be extended to any linear combination of the two, thereby restoring generality to our line of reasoning and bringing Eq. (B.15) into exactly the form of Eq. (B.12).The remainder of the argument given for the ground-state case is thus exactly applicable here, and the theorem is proved.

References

1. Dover, C.B., and Lemmer, R.H., *Phys. Rev.*, **165,** 1105 (1968).
2. *The Many-Body Problem,* Univeristé de Grenoble, École d'été de physique théorique, Les Houches, 1958 (Dunod–Wiley, New York, 1959); lectures.
3. Pines, D., *The Many-Body Problem* (Benjamin, New York, 1961); lectures and reprints.
4. Thouless, D.J., *The Quantum Mechanics of Many-Body Systems* (Academic Press, New York, 1961).
5. Van Hove, L., Hugenholtz, N.M., and Howland, L.P., *Quantum Theory of Many-Particle Systems* (Benjamin, New York, 1961); reprints.
6. Kadanoff, L., and Baym, G., *Quantum Statistical Mechanics* (Benjamin, New York, 1962).
7. Morrison, H.L. (Ed.), *The Quantum Theory of Many-particle Systems* (Gordon and Breach, New York, 1962); reprints.
8. Bonch-Bruevich, V.L., and Tyablikov, S.V., *The Green Function Method in Statistical Mechanics,* transl. D. ter Haar (Interscience, New York, 1962).
9. Abrikosov, A.A., Gorkov, L.P., and Dzyaloshinskii, I.E., *Methods of Quantum Field Theory in Statistical Physics,* transl. R.A. Silverman (Prentice-Hall, New York, 1963).
10. Schultz, T.D., *Quantum Field Theory and the Many-body Problem* (Gordon and Breach, New York, 1963).
11. Percus, J.K. (Ed.), *The Many-body Problem* (Interscience, New York, 1965); Proceedings of the Symposium on the Many-body Problem, Stevens Institute of Technology, January, 1957.
12. Nozières, P., *Theory of Interacting Fermi Systems,* transl. D. Hone (Benjamin, New York, 1964).
13. Kirzhnits, D.A., *Field Theoretical Methods in Many-Body Systems,* transl. A.J. Meadows (Pergamon, Oxford, 1967).
14. Mattuck, R.D., *A Guide to Feynman Diagrams in the Many-body Problem* (McGraw-Hill, New York, 1967).

15. Landau, L. D., *Soviet Phys.–JETP*, **3,** 920 (1957).
16. Bardeen, J., Cooper, L. N., and Schrieffer, J. R., *Phys. Rev.*, **108,** 1175 (1957).
17. Bogoliubov, N. N., Tolmachev, V. V., and Shirkov, D. V., *A New Method in the Theory of Superconductivity* (Consultants Bureau, New York, 1959).
18. Bogoliubov, N. N., *J. Phys. (U.S.S.R.)*, **11,** 23 (1947).
19. Gor'kov, L. P., *Soviet Phys.–JETP*, **7,** 505 (1958).
20. Gell-Mann, M., and Brueckner, K. A., *Phys. Rev.*, **106,** 364 (1957).
21. Kadanoff, L. P., and Martin, P. C., *Phys. Rev.,* **124**, 670 (1961).
22. Balian, R., Bloch, C., and De Dominicis, C., *Compt. Rend.*, **250,** 2850 (1960).
23. Nozières, P., and Pines, D., *Nuovo Cimento*, **9,** 470 (1958) (Ser. 10).
24. Jastrow, R., *Phys. Rev.*, **98,** 1479 (1955).
25. Ayres, R. U., *Phys. Rev.*, **111,** 1453 (1958).
26. Matsubara, T., *Progr. Theoret. Phys.*, **14,** 351 (1955). Matsubara's proof of the thermodynamic Wick's theorem is incomplete; Thouless [54] gives the first complete proof.
27. Goldstone, J., *Proc. Roy. Soc. (London), Ser. A*, **239,** 267 (1957).
28. Hugenholtz, N. M., *Physica*, **23,** 481 (1957).
29. Montroll, E. W., and Ward, J. C., *Phys. Fluids*, **1,** 55 (1958).
30. Bloch, C., and De Dominicis, C., *Nuclear Phys.*, **7,** 459 (1958); **10,** 181, (1958).
31. Glassgold, A. E., Heckrotte, W., and Watson, K. M., *Phys., Rev.*, **115,** 1374 (1959).
32. Luttinger, J. M., and Ward, J. C., *Phys. Rev.*, **118,** 1417 (1960).
33. Lee, T. D., and Yang, C. N., *Phys. Rev.*, **113,** 1165 (1959); **116,** 25 (1959); **117,** 12 (1960); **117,** 22 (1960).
34. Klein, A., and Prange, R., *Phys. Rev.*, **112,** 994 (1958).
35. Martin, P. C., and Schwinger, J., *Phys. Rev.*, **115,** 1342 (1959).
36. Luttinger, J. M., *Phys. Rev.*, **121,** 942 (1961).
37. Konstantinov, O. V., and Perel', V. I., *Soviet Phys.*, **12,** 142 (1961).
38. Alekseev, A. I., *Soviet Phys.–Uspekhi*, **4,** 23 (1961). Review article; good references to previous Soviet literature.
39. Beliaev, S. T., *Soviet Phys.–JETP*, **7,** 289 (1958).
40. Hugenholtz, N. M., and Pines, D., *Phys. Rev.*, **116,** 489 (1959).
41. Kromminga, A. J., and Bolsterli, M., *Phys. Rev.*, **128,** 2887 (1962).
42. Wentzel, G., *Quantum Theory of Fields* (Interscience, New York, 1949).

43. Schweber, S., *An Introduction to Relativistic Quantum Field Theory* (Row, Peterson, Evanston, 1961), Chap. 6.
44. Dirac, P. A. M., *Quantum Mechanics*, 4th Ed. (Oxford University Press, Oxford, 1958).
45. Fisher, J. C., *Phys. Rev.*, **125,** 492 (1962).
46. Fröhlich, H., *Phys. Rev.*, **79,** 845 (1950); *Proc. Roy. Soc. (London), Ser. A*, **215,** 291 (1952).
47. Hugenholtz, N. M., and Van Hove, L., *Physica*, **24,** 363 (1958).
48. Lighthill, M. J., *Fourier Analysis and Generalized Functions* (Cambridge University Press, Cambridge, 1958).
49. Bremermann, H. J., and Durand, L., *J. Math. Phys.*, **2,** 240 (1961).
50. Wick, G., *Phys. Rev.*, **80,** 268 (1950).
51. Dyson, F. J., *Phys. Rev.*, **75,** 486 (1949).
52. Gell-Mann, M., and Low, F., *Phys. Rev.*, **84,** 350 (1951), Appendix.
53. Baker, G. A., *Phys. Rev.*, **131,** 1869 (1963).
54. Thouless, D. J., *Phys. Rev.*, **107,** 1162 (1957).
55. Dzyaloshinskii, I. E., *Soviet Phys. -JETP*, **15,** 778 (1962).
56. Baym, G., and Sessler, A. M., *Phys. Rev.*, **131,** 2345 (1963).
57. Baym, G., and Mermin, N. D., *J. Math. Phys.*, **2,** 232 (1961).
58. Craig, R. A., *J. Math. Phys.*, **9,** 605 (1968). See also Schwinger, J., *J. Math. Phys.*, **2,** 407 (1961), for the application of a similar technique in a different context.
59. Gaudin, M., *Nuclear Phys.*, **15,** 89 (1960).

Index